SPACE
THE FIRST 50 YEARS

USA

SPACE

THE FIRST 50 YEARS

SIR PATRICK MOORE AND H J P ARNOLD

FOREWORD BY BUZZ ALDRIN

STERLING

New York / London

www.sterlingpublishing.com

To Douglas—a great photographer, a great author, and above all, a great friend.

Space: The First 50 Years

Library of Congress Cataloging-in-Publication Data Available

1 2 3 4 5 6 7 8 9 10

Published in 2007 by Sterling Publishing Co., Inc., 387 Park Avenue South, New York, NY 10016

Original published in the UK in 2007 by Mitchell Beazley, an imprint of
Octopus Publishing Group, 2–4 Heron Quays London E14 4JP, England

Distributed in Canada by Sterling Publishing c/o Canadian Manda Group,
165 Dufferin Street Toronto, Ontario, Canada M6K 3H6

For information about custom editions, special sales, premium and corporate purchases, please
contact Sterling Special Sales Department at 800-805-5489 or specialsales@sterlingpub.com

Color reproduction by Bright Arts in China

Printed and bound by Toppan Printing Company in China

Sterling ISBN-13: 978-1-4027-5208-7
ISBN-10: 1-4027-5208-3

contents

FOREWORD

Buzz Aldrin

Discovery never ends.

This book marks fifty years since the launch of Sputnik 1, the first artificial satellite to be sent into space. Fifty years since mankind's reach extended beyond its own planet. Fifty years of discovery.

When stepping out of the *Apollo 11* lunar module with Neil Armstrong, the first men to stand on the moon, I was struck immediately by the moon's strange horizon. It visibly curved, like nowhere on Earth. It was extremely disorientating.

"Magnificent desolation" are the words I have most often used to describe that remarkable location. No other description comes close.

Sir Patrick Moore is never stuck for words, as *Space* consistently demonstrates. His boundless enthusiasm for astronomy is as intelligent as it is infectious. Traveling with his words is like taking a museum tour of the universe, hosted by a favorite teacher.

H.J.P. Arnold—to whom this, his final work, is a tribute—safeguarded our imagination. His expertise with astronomical photography is unmatched, his contribution to our Apollo photo record immeasurable.

Space exploration has produced some of the most iconic images of all time, and Arnold's collection of the fifty greatest space images cannot help but fascinate. I am proud, and a little humbled, to have taken the image in first place—a simple footprint.

"Most iconic" is the mark of boot on a surface unlike any found at home. Not a rocket, dazzling in its technology and power; not some beautiful, distant nebula; but a simple sign of man's arrival on a surface beyond the earth. Evidence that we have walked further. A step into that magnificent desolation.

What will mark the first stages of the next fifty years of adventure? Where will we travel next? To the moon again, of course, and on to Mars. The red planet marks a new and exciting challenge. To leave another print on the surface of an entire other world.

Closer to home, commercial space travel is achievable within our lifetimes. Affordable journeys into space for anyone, everyone. A chance to inspire a new generation.

As a child I didn't always dream of going into space. As an adult, my dreams have been there often. You don't always know where the journey will take you. How lucky I have been to travel so far, to live a dream, and to take so many imaginations with me.

We need books like the one you hold in your hands. For inspiration, for the next curved horizon, for the journey.

Because discovery never ends.

Opposite: One of the most detailed true-color images ever obtained of the earth. Based on a collection of satellite observations, the view is essentially a mosaic of every square mile of our planet. The earth is the only planet in the solar system with vast oceans of liquid water on its surface—the result being that at any given time more than half of our planet is covered by water-bearing clouds.

INTRODUCTION

Sir Patrick Moore

On October 4, 1957 the space age began—not with a whimper, but with a very pronounced bang. Russia's artificial satellite Sputnik 1 blasted off, and sped round the world transmitting the "bleep bleep" signals that will never be forgotten by anyone who tuned into them. It was one of the great moments in human history, and will be remembered long after the Battle of Hastings and even WWII have faded from memory.

The idea of space travel was not new. As long ago as the second century AD, a Greek satirist, Lucian of Samosata, wrote a story about a voyage to the moon, calling it the "True History" (though admittedly he was careful enough to point out that it was made up of nothing but lies from beginning to end). But it is only in our own time that reaching other worlds has become possible. Little more than half a century ago, anyone who talked seriously about reaching the moon was regarded, at best, as eccentric. One eminent astronomer went so far as to describe it as "utter bilge."

Since 1957 we have achieved much. In 1961 Yuri Gagarin became the first astronaut; little more than eight years later Neil Armstrong stepped out on to the bleak rocks of the lunar Sea of Tranquillity. Space stations have been set up, and unmanned probes have bypassed all the planets at reasonably close range. Controlled landings have even been made on two of them, Mars and Venus, as well as the tiny asteroid Eros. Europe's probe *Giotto* penetrated the head of Halley's comet, and samples have been brought back from another comet, Wild 2.

This book is not intended to be a history of astronomy, or even a proper description of the astronomical scene, but it does seem useful to say a little about what has happened in past ages. After all, astronomy is one of the oldest sciences in the world; even our remote cave-dwelling ancestors must have looked up at the skies and marvelled at what they saw there. So let us spend a few moments in looking back into the earlier days.

I have divided the story up into a few definite "eras": Many people will disagree, and it is hard to be strictly chronological because there is considerable overlap. However, the procedure may be helpful. So here are my eras:

Left: Gurkhani Zij observatory in Samarkand, Uzbekistan. Ulugbek, or Ulugh Beg, who built this observatory in 1428, used it to catalog the positions of nearly 1,000 stars.

Below: The Copernican World System. Nicolaus Copernicus was the first European astronomer to formulate the modern heliocentric theory of the solar system, arguing that the sun and not the earth was the center of the universe.

1. From the earliest times to the first four ivilizations—up until the reign of Menes, the first gyptian monarch, in around BC 3100. The era of nythology and astrology.

2. The 4th millennium BC to the start of the Greek eriod. Systematic observations made by the gyptians, the Chinese, and the Babylonians. The earth was thought to be flat and to lie in the center of the universe, but at least some accurate measurements were made, and good calendars were produced.

3. The start of true science, by the Greeks. The irst great Greek astronomer was Thales of Milstus born c.BC 624) and the last was Ptolemy of Alexandria died c. 180 AD). The earth was found to be a globe ather than a flat plane. Planets visible to the naked ye were identified, and their movements worked out. henomena such as eclipses were understood. Good tar catalogs were compiled. The size of the earth was neasured with remarkable accuracy, and a few stronomers, notably Aristarchus (c 310 BC–c 230 BC), proposed that the sun, not the earth, was the center o the known Universe. Ptolemy also drew a map of the civilized world which was based upon astronomica observations; it even showed the British Isles.

4. Interim period, from Ptolemy's time to 1543 AD. Very little progress was made until the rise o Arabian astronomy with leaders such as Al-wa'mur who founded the Baghdad school in 813 AD. Better sta catalogs and planetary tables were compiled, an there were major observatories—notably that of Ulugh Beg (more properly Ulugbek), established a Samarkand in 1433. In 1449 Ulugh Beg wa assassinated on the orders of his son, whom he ha banished on astrological advice, and subsequen developments were mainly European.

5. The "Great Revolution," 1543–1689. The theory that the earth is a planet moving around the su was given in a book by a Polish churchman, Copernicu (Nicolaus Kopernik), in 1643. It met with grea opposition, mainly from the Catholic church—indeed

Giordano Bruno was burned at the stake in Rome in 1600, partly because of his support for the Copernican theory as opposed to the Ptolomatic. The "Revolution" was ended in 1689 with the publication of Newton's *Principia* sometimes described as the greatest mental effort ever made by one man. The seventeenth century was also the beginning of the telescopic era. The first astronomical telescopes were brought into use around 1610, mainly by Galileo, who used his tiny "optick tube" to make a series of spectacular discoveries such as the craters of the moon, the phases of Venus, the satellites of Jupiter, and the countless stars of the Milky Way. He was a strong supporter of the Copernican theory; in 1633 he was brought to trial by the Inquisition, threatened with torture, and forced to "curse, abjure and detest" the false concept that the earth moves around the sun. He was finally pardoned in 1992.

6. The Instrumental Era, 1689–1906. The advent of larger telescopes William Herschel's 49 inch (124.5 cm) in 1789, and Lord Rosse's 72 inch (182.9 cm) in 1846. The first star distance measured by F. W. Basel in 1838 giving a much improved idea of the scale of the universe. Astronomica

to sum up the state of our astronomical knowledge in 1957.

The solar system is our home in the universe. It is made up of one star (the sun), the eight planets, and various lesser bodies such as asteroids and comets. Most of the planets have satellites; we have one (our faithful moon) but Jupiter has over 60, although only four of them are large. The system is divided into two obvious parts. First come four "terrestrial" planets—Mercury, Venus, the earth, and Mars—of which the earth is the largest, though in size and mass it is almost a twin of Venus. Beyond the orbit of Mars there is a wide gap, in which move many thousands of small bodies known as

Right: The floodlit 250 ft. (76 m) diameter dish of the Mark 1A radio telescope. Commissioned in 1957, the Mark 1 was the world's first giant dish radio telescope, designated the Mark 1A after improvements in the 1970s. It was officially renamed the Lovell telescope in 1987, in honor of Bernard Lovell, who founded the Jodrell Bank Observatory.

to that of the earth, so that if you draw a plan of the solar system on a flat piece of paper you are not very far wrong. That is why the planets keep to the Zodiac instead of wandering all over the sky.

Of course, the moon is the closest of all the celestial bodies, and the only one that moves round the earth. Its mean distance is rather less than 250,000 miles (400,000 km), and it takes 27.3 days to complete one orbit. Since the sun can light up only half of it at any one time, the moon shows regular phases, or changes of shape, from new to full. When the moon lies between the sun and the earth its dark "night" side is turned towards us, and the moon cannot be seen—unless the alignment is exact, and the new moon passes directly in front of the sun, briefly hiding it and causing a solar eclipse. As the moon moves along it appears as a slender crescent in the evening sky, waxing to a half, and then full, after which the phase decreases again and the moon returns to new. It is often thought that moonlight can sometimes be nearly as bright as daylight. In fact, it would take about half a million full moons to rival the light of the sun!

Comets are the most erratic members of the sun's family. They are not solid and rocky, like planets; a comet has been described as a "dirty iceball." Although they move round the sun, most of them do so in very eccentric orbits. Hundreds of comets have short orbital periods, so that we can keep track of them, but all these are faint with the exception of Halley's comet; which has a period of 76 years and will next return to perihelion in 2061. Really brilliant comets have much longer periods so that we never know when to expect them. Many people will remember the lovely comet Hale-Bopp, which hung in the sky for months from the summer of 1996 to the early autumn of 1997. Alas, it will not be seen again for four thousand years.

As a comet flies, it leaves a dusty trail in its orbit. When one of these particles crashes into the earth's upper air, it is heated by friction against the air particles, and burns away to produce the luminous streak which we call a shooting star or meteor. Now and then a larger body arrives on earth, and is then called a meteorite—but there is no connection between meteorites and shooting star meteors; meteorites come from the asteroid belt. They may make craters; one, in Arizona, is nearly 1 mile (1.6 km) wide. The moon is covered with meteorite craters, and I will have more to say about them in Chapter 3 (see pages 40–9).

The sun rules the solar system. It is a huge globe of incandescent gas, 865,000 miles (1,392,083 km) in diameter; its volume is over a million times that of earth. Without it, earth and the other planets would never have come into existence; they were formed in a cloud of material surrounding the sun about 4,600,000,000 years ago.

The sun has a surface temperature of about 10,832°F (6,000°C), but it is not burning in the manner of a coal fire; a star made up of coal and burning as fiercely as the sun does would turn to ashes in a few million years—but the sun is certainly older than the earth. Its energy is due to nuclear reactions going on deep inside it. The most plentiful element in the universe is hydrogen, and the sun contains a great deal of it. Near the core, where the pressure is colossal and the temperature is around 27,000,000°F (15,000,000°C), the nuclei of hydrogen atoms are combining to make up nuclei of a heavier element, helium. It takes four "bits" of hydrogen to make one "bit" of helium; every time this happens, a little energy is set free and a little mass (or weight) is lost. It is this energy which keeps the sun shining. The mass loss amounts to 4 million metric tons per second, so the sun weighs much less now than it did when you opened this book. Eventually, the sun will run short of hydrogen fuel, and will change, but there is no immediate cause for alarm; nothing much will happen for at least a thousand million years yet.

Now let us look further, way beyond our solar system. The sun is a normal star, but at its distance of 93,000,000 miles (149,668,992 km) from the earth it is a very near neighbor on the astronomical scale. It is one of

Opposite: An artist's impression of our solar system. From left to right: the Sun, Mercury, Venus, Earth, Mars, Jupiter, Saturn, Uranus, and Neptune. Note that Pluto is not included; in 2006 it was downgraded from a planet to a Kuiper Belt object.

Right: This image, depicting part of the sunspot group near the disk center, was taken by the Swedish 1-m Solar Telescope in May 2002.

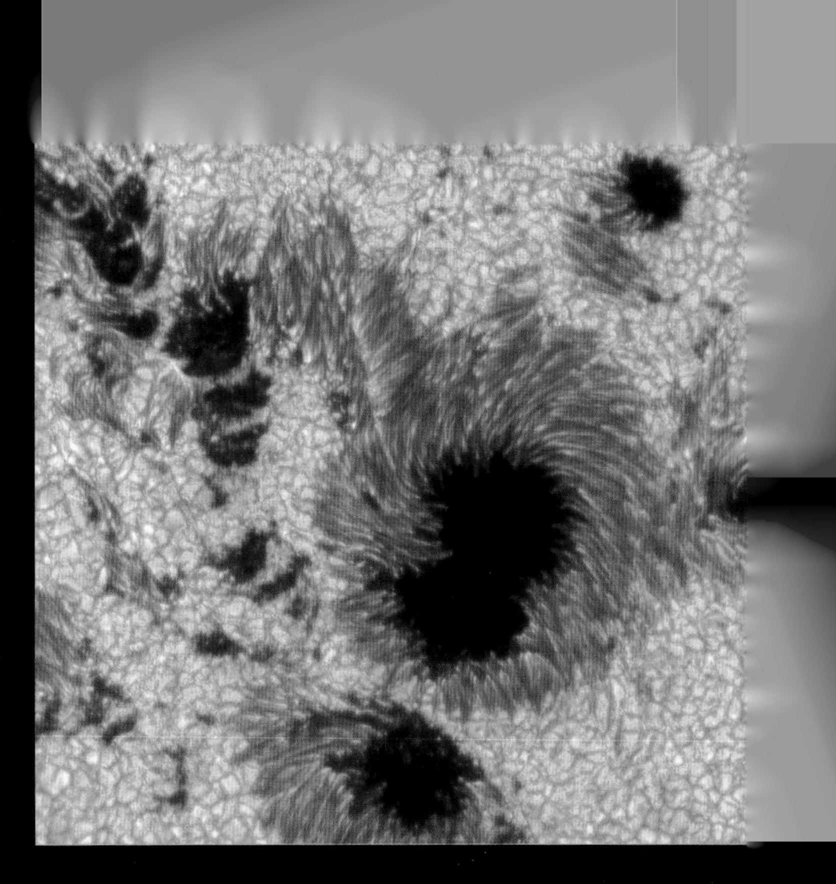

about 100,000 million suns making up our star system or galaxy. The stars are so far away that we give their distances not in miles or kilometers but in light-years. Light does not travel instantaneously; it flashes along at 185,000 miles per second (297,728 kilometers per second), so that in a year it covers 5,800,000 miles (9,334,195 km)— and this is what is known as a light-year. If you want to turn miles into light-years, multiply by 6 million million, which will give you a good approximation. The nearest star beyond the sun is 4.21 light-years away—roughly 25 million million miles. The universe is truly a large place!

The stars are not fixed in space; they are moving in all sorts of directions at all sorts of speeds, but they are so far away from us that their individual or "proper" motions are very slight. To all intents and purposes the constellations look the same now as they must have done to King Canute and Julius Caesar, or the builders of Stonehenge. It is only our nearest neighbors, the members of the solar system, which wander around. The constellation patterns we use were those drawn up by the Greeks; Ptolemy gave a list of 48, all of which are still in use. If we used, for example, the Chinese or the Egyptian patterns, our star maps would look very different though of course the stars would be exactly the same.

In fact, the constellations have no significance whatsoever, because the stars in them are at different distances from us and have no real connection with each other.

The galaxy is a flattened system, with the solar system more or less in the main plane, but about 26,000 light-years from the center. When we look along the main plane, we see many stars in the same direction; this produces the effect of the lovely band we call the Milky Way. The stars there look as if they are crowded together, but this is not so; they are widely spaced. Collisions must be rare.

The stars are of various types; some are hotter than the sun, some cooler. The real "cosmic searchlights" shine well over a million times as powerfully as the sun, but we also find stellar glow worms, shining very feebly. There is endless variety in the galaxy.

We can also see other galaxies, millions of light-years from us—in fact the latest surveys indicate that at least 1,000 million galaxies are within range of our telescopes. We can now reach out to over 12,000 million light-years. So the total number of stars known to us is unimaginably great, and there is no doubt that many of these stars have planetary systems of their own.

It is reasonable to believe that Earth-like planets may support Earth-like life, but we have no proof of this. If other civilizations exist we may, at some time, be able to make radio contact with them, but travel to other star systems is quite beyond anything we can plan as yet. We must start closer to home in our own solar system.

Top left: **1950**
The Korean War begins on June 25 with North Korea's invasion of South Korea. U.S. troops respond on October 7, 1950 with the invasion and capture of the port of Inchon. The war ends inconclusively in July 1953, with an armistice between the UN, U.S., China, and North Korea. South Korea refuses to sign the armistice, leaving the two Koreas separate. The war establishes a precedent for U.S. intervention to contain Communist expansion.

Bottom left: **1955**
April 12 – Following 8 years of painstaking research, American physician Dr. Jonas Salk announces his discovery of the polio vaccine at the University of Pittsburgh Medical School. Salk, hailed as a miracle worker, seeks neither wealth nor fame through his innovations and refuses to patent the vaccine.

Top right: **1953**
March 5 – Soviet Communist leader Joseph Stalin dies. Hundreds of thousands of people line up to see his body lying in state at the hall of Trade Union House, Moscow.

Bottom right: **1956**
Elvis Presley's legendary breakthrough performance on *The Ed Sullivan Show* takes place on September 9. The show is viewed by a record 60 million people—82.6 percent of the U.S. television audience and the largest single audience in television history.

1950–1959

Top: **1957**
On March 25, European foreign affairs ministers meet in Rome to sign the treaties creating the European Economic Community. The EEC creates a common market, eliminating barriers to the movement of goods, services, capital, and labor, and imposes a common agricultural policy (CAP) and a common external trade policy.

Bottom left: **1958**
Mao Tse-tung, Chairman of the People's Republic of China and the chief architect of the Chinese Revolution, launches the Great Leap Forward, an economic and social campaign to increase China's industrial production by mobilizing the country's enormous manpower into rural peoples' communes.

Bottom right: **1959**
February 16 – Cuban revolutionary leader Fidel Castro is sworn in as Prime Minister of Cuba.

Opposite: Following the end of the war, von Braun and most of the key members of the German rocket team went to America, first to the White Sands proving ground in New Mexico. For the next few years they adapted captured V2s (as shown here) for scientific use, and laid the foundations of the whole U.S. space program.

01 INTO SPACE

t was originally thought that space travel was a concept belonging to science fiction. Lucian of Samosata set the cene, around 1500 AD; his *True History* is worth reading even today, although his methods would hardly appea o NASA. In the story, a ship sailing happily through the Pillars of Hercules (known to us as the Straits of Gibraltar was caught in a waterspout, and hurled upward so violently that after seven days and seven nights it landed or he Moon. The sailors found that they had arrived at an opportune moment; the King of the Moon was about tc do battle with the King of the Sun, following an argument about which of the monarchs should have first rights or he planet Venus. Other stories followed. Before coming to the ways in which we can cross space, let us dispose of the ways in which we cannot.

Consider the *Somnium*, written by the great astronomer Johannes Kepler; it was published (posthumously) ir 1634. Kepler's hero is carried to the moon by obliging demons. Five years later came Bishop Godwin's *Man in the Moone*, describing how the narrator, Domingo Gonzales, trains wild geese to tow him through the air on a raft— only to discover, to his alarm, that the birds hibernated on the moon and were setting off for their winter quarters taking him with them. Cyrano de Bergerac, in 1657, had a different idea. He reasoned that since dew on a grass awn disappears after dawn, the sun must suck it up, so that all you have to do is collect a large number of bottles ill them with dew, fasten them around yourself, stand out on a fine morning and let the sun do the rest!

The first space travel novel with a scientific basis was written in 1860 by Jules Verne: *From the Earth to the Moor* the sequel, *Round the Moon*, followed a few years later). Three intrepid travelers—Barbicane, Captain Nochol, anc Michel Ardan—enter a projectile and are fired moonward from the barrel of a powerful cannon blasting away at a speed of 7 miles per second (11.27 km per second). Verne was not a qualified scientist, but he believed in keeping o the facts as well as he could, and was remarkably successful. He was correct in saying that the journey woulc have to start at a speed of 7 miles per second (11.27 km per second), about 25,000 mph (40,000 km/h).

Throw an object upward, and it will rise, pause, and then fall back to the ground. Throw it faster, and it will rise higher before returning. If you could hurl it up at 7 miles per second (11.27 km per second), it would never come down at all; the earth's pull of gravity would not be strong enough to hold it, and the object would escape into space. This is why 7 miles per second (11.27 km per second) is known as the earth's "escape velocity." At first sight, therefore, the spaceman idea seems reasonable.

Unfortunately, there were two points which Verne did not take into account. The first is that air sets up resistance by friction—as you can see whenever you pump up a bicycle tire, a projectile moving through the dense lower atmosphere at escape velocity would burn away even before it reached the mouth of the cannon. Secondly, starting off at this speed would be quite a jerk, to put it mildly, and the luckless travellers would at once be turned to jelly. The other great science fiction pioneer, H. G. Wells, planned an anti-gravity device, but we have no idea how this could be achieved, and quite probably it can never be achieved at all.

ROCKET SCIENCE

Space guns, anti-gravity, and the like having failed us, we must look elsewhere. At the moment there is only one answer: the rocket, which works upon what Newton called the Principle of Reaction—"every action has an equal and opposite reaction."

Consider a firework rocket. It is made up of a hollow tube filled with gunpowder. When you light the fuse, the gunpowder starts to burn, and hot gas is produced. The gas streams out through the exhaust and, as it does so, pushes the tube in the other direction. In effect, the rocket is "pushing against itself," and continues to fly so long as the gas keeps on streaming out. There is no need for surrounding air—in fact air is a nuisance, because it sets up resistance. The rocket is at its most efficient in the vacuum of space.

Rockets were invented many centuries ago, probably by the Chinese, and were used for firework displays, as indeed they still are. Inevitably they were also used in warfare and when the Mongols attacked the city of Kai-Fung-Fu in 1232, the Chinese defenders retaliated with the "arrows of flying fire." Much later, several European nations formed rocket corps, with a certain amount of success. And in 1814, during the war between Britain and the United States, a small British force fired missiles against American troops at Bladensburg—hence the famous line about "the rockets' red glare" which is part of the modern American national anthem.

Nobody in their right mind would even think about using gunpowder rockets for space travel, though it is on record that in the year 1500 or thereabouts a Chinese merchant named Wan-Hoo tried to launch himself in that way. He fastened 47 rockets on to himself and ordered his servants to light them all at once—which they did, with predictable results. However, for real developments we must move on to the last decade of the nineteenth century, with the work of a Russian named Konstantin Eduardovitch Tskolkovskii. In 1895 he wrote a novel called *Beyond the Planet Earth.* It was not translated into English until 1960, and as a literary effort it can only be described as atrocious, but as a scientific forecast it was many years ahead of its time.

Tskolkovskii grasped several vitally important points. Firstly, rockets alone can work in airless space, but fuels such as gunpowder are useless, if only because they cannot be controlled after they have been ignited. Secondly, friction against the air means starting off relatively slowly, and accelerating to escape velocity only after the thick lower atmosphere has been left behind. Therefore he proposed to use liquid propellants, such as petrol and liquid oxygen, which would be forced into a combustion chamber by means of a pump. They would react together, gas would be produced; the gas would stream out of the exhaust—and the rocket would fly, gradually at first and then with ever-increasing speed.

A liquid-propellant launcher would be much more powerful than gunpowder, but still not good enough to work up to escape velocity, so Tskolkovskii planned to use a step vehicle. In this arrangement a small rocket is mounted on top of a large one. At first the lower stage does all the work; when it has used up all its fuel it falls away, leaving the upper stage to continue the journey under its own power. In effect, the upper step has been given a "running jump" into space.

Tskolkovskii's first really important scientific paper was published in 1905. It caused virtually no interest, partly because Tskolkovskii was only an obscure schoolteacher and partly because the article was published in a low-circulation Russian journal. Moreover, Tskolkovskii was purely a theorist, and never personally fired a rocket in his life.

Robert Hutchings Goddard, in the United States, knew nothing about Tskolkovskii's paper, but he was very interested in rockets. A professional scientist at Clark University, he had plenty of opportunity to carry out practical experiments. His main aim was to investigate the conditions in the upper atmosphere, but in a paper published in 1920 he also suggested that it should be possible to send a small rocket to the moon. This idea was ridiculed in some sections of the press, and Goddard was so annoyed that thereafter he avoided publicity—with the result that when he launched the first liquid-propellant rocket in history, in 1926, very few people heard about it until much later. The rocket itself was modest, travelling a mere 184 ft. (56 m) at a maximum speed of 60 mph (95 km/h), but it was the ancestor of the huge interplanetary rockets of today.

Meanwhile, in Europe, interest had been aroused by the publication of a book by a Romanian mathematician, Hermann Oberth.

It was the first more or less popular book on space research (although it did contain some reasonably complicated calculations), and, surprisingly, it became something of a bestseller. In Germany a group of experimenters, one of whom was Wernher von Braun, set up an elaborate testing ground and began serious work. The results were promising, and before long there were rockets capable of rising to over 1 mile (1.6 km). The "Rocket Flying Field" near Berlin flourished, but then politics intervened; the Nazis came to power and looked hard at the possibility of using rockets in war. The experiments were transferred to Peenemünde, a village on the small Baltic island of Usedom, and kept themselves very much to themselves. By the autumn of 1939, when Hitler was ready to attack, von Braun and his team had built a rocket over 20 feet (6 meters) long, able to soar to over 5 miles (8 km) above the ground—almost to "the edge of space." Oberth went to Peenemünde, but Tskolkovskii and Goddard did not.

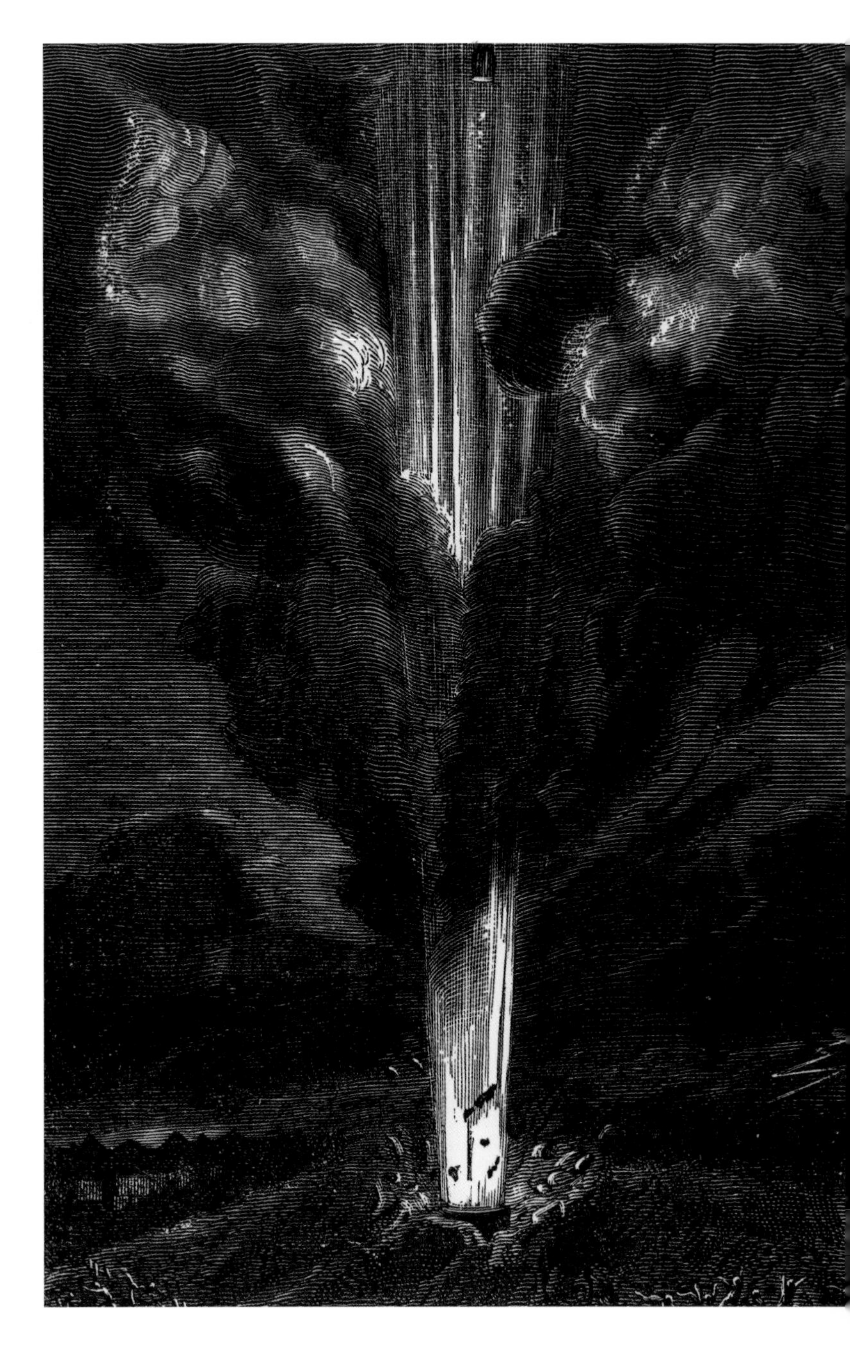

Above: In 1886 the great French novelist Jules Verne published *From the Earth to the Moon*. This old woodcut from the first edition of the book shows the space travellers in a capsule that is launched from a cannon. The idea is impractical for many reasons, but Verne did give the right starting velocity: 7 miles per second (11.27 km per second).

THE V2

By 1942 there was a new rocket, the A4, later to become famous (or notorious!) as the V2, used to bomb London and Norwich in the last stage of the war. This was not the experimenters' original aim. When the first V2 was tested successfully, von Braun commented: "A good flight—but the payload landed on the wrong planet!"

When the Russians were advancing rapidly and it was clear that Germany had lost the war, von Braun and most of his team made their way out of Peenemünde and came over to the side of the Allies, leaving the Soviet forces to seize what had been left behind. Before long the German scientists were working in the United States, this time for peace rather than war.

The German scientists and the Americans were certainly energetic. Captured V2s were used as test vehicles; step rockets were built; scientific payloads were sent aloft, and there was unceasing activity at the proving grounds, first at White Sands in New Mexico and then at Cape Canaveral in Florida. Then, on July 29, 1955, the artificial satellites program was officially approved, with practical experiments to take place at some time between July 1957 and December 1958. This was the time of the International Geophysical Year, when scientists from all countries were due to combine in studies of the earth in all its aspects. (Actually the IGY lasted for 18 months.)

Popular interest was immediate; suddenly people realized that space missions were not only possible, but were imminent. Looking back, it is clear that an artificial satellite could have been launched in 1956 or even earlier, but politics intervened. There were several competing programs, and the man who knew more about rocketry than anyone else—Wernher von Braun—was German, though by this time he had become an honorary citizen of the United States. What

America did not realize was that the Russians had been far from idle. Few of the Peenemünde team had gone to the USSR, but the Russians had an outstanding researcher of their own: Sergei Korolev, officially referred to as the "Chief Designer." He had had a checkered career, and had even spent some time in one of Stalin's concentration camps, but by 1957 he had become the undisputed hero of the Soviet rocket program. He and von Braun always wanted to meet, but they never did.

SPUTNIK 1 AND THE SPACE RACE

Sputnik 1 was launched by a step vehicle, just as Tskolkovskii would have expected. It was put into an orbit which took it from 135 to 580 miles (215 to 935 km) above ground level, and it completed one circuit of the earth in just over 90 minutes. It was spherical, with a diameter of less than 2 ft. (60 cm), and it carried little apart from its radio transmitter. It was too faint to be seen with the naked eye, but the uppermost stage of the carrier rocket was brighter, and was observed by watchers all over the world. Its signals were easy to pick up, and the vehicle itself was tracked both in the USSR and by the then-new 250 ft. (75 m) radio telescope at Jodrell Bank in Cheshire, now known as the Lovell Telescope in honor of Sir Bernard Lovell, but for whom it would not have been built.

Sputnik 1 moved in the same way as a natural astronomical body would have done, but it was not completely above the upper atmosphere and was therefore affected by air drag. Gradually it was dragged down, and in January 1958 it reentered the lower layers of the atmosphere and burned away in the manner of a shooting star meteor. Only above a height of about 150 miles (250 km) is an orbiting vehicle really safe from the effects of friction.

It cannot be said that reactions in the West to Sputnik 1 were wholly favorable. The Americans were alarmed at having fallen behind in what was already being called "the space race," and one disgruntled general went so far as to describe Sputnik 1 as "a hunk of old iron that almost anybody could launch." When the Soviets sent up the dog-carrying Sputnik 2, on April 14, 1958, the United States authorities swallowed their pride and turned to von Braun.

Opposite: Robert Hutchings Goddard, the American pioneer who launched the first liquid-propellant rocket. It was modest, but it was of immense importance. Goddard also speculated about the possibility of sending a projectile to the moon. For this he was ridiculed by the popular press, and was therefore reluctant to publish his work.

Below left: A Russian engineer checks Sputnik 1, the first artificial satellite, launched on October 4, 1957. It carried little apart from the transmitter which sent out the famous "Bleep! Bleep!" signals, but it marked the real beginning of the space age.

Below right: From left to right: Dr William H Pickering, Director of the Jet Propulsion Laboratory, Dr James Van Allen, a scientist from University of Iowa, and Wernher von Braun, one of the leading figures in the development of rocket technology, raise a full-size model of America's first satellite, Explorer 1, above their heads following a successful launch on January 31, 1958.

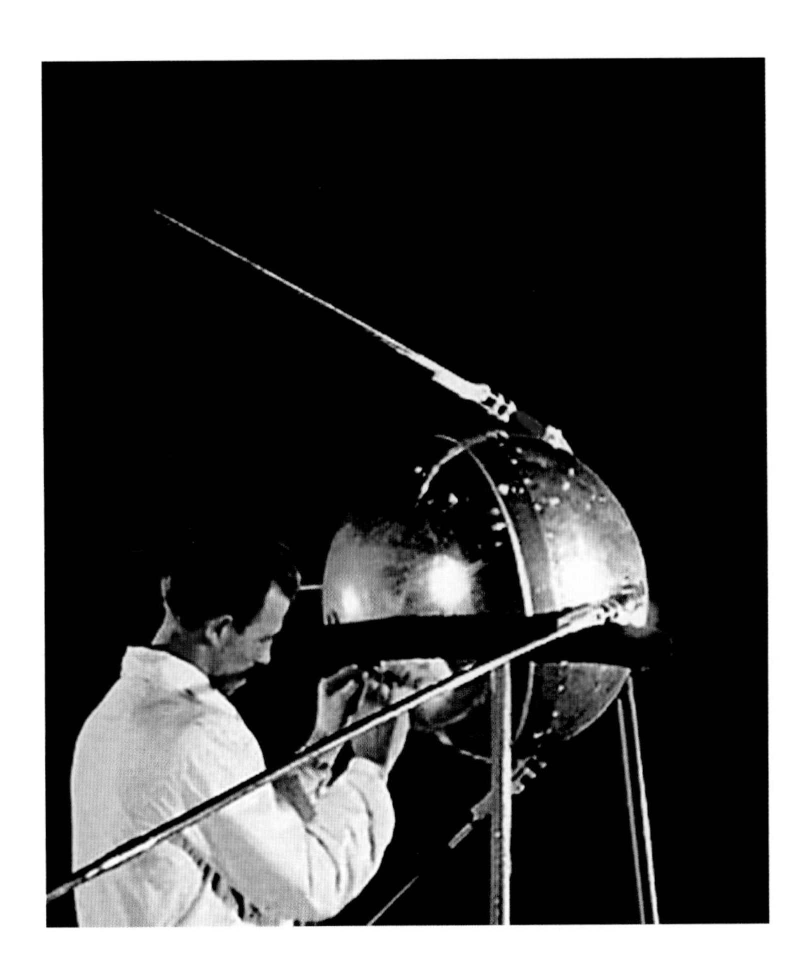

THE VAN ALLEN BELT

It took von Braun only three months to launch America's first satellite, Explorer 1—which made the first major discovery of the satellite era. Instruments in it detected the belts of radiation surrounding the earth which are now known as the Van Allen Belts, because the equipment used was designed by the American scientist James Van Allen.

This discovery was fortuitous. Explorer 1's orbit was decidedly elliptical, with its perigee (nearest point) at 224 miles (360 km) and apogee (furthest point) at 1,585 miles (2,550 km); the period was 115 minutes. One of its instruments was a Geiger counter, to detect charged particles and high-energy radiation; it was hoped to investigate the numbers of cosmic rays at various heights above the ground. Cosmic rays are not rays at all but atomic nuclei coming from space. The cosmic ray counts were expected to increase with altitude, because our upper air breaks up the comparatively massive "primaries" and prevents them from reaching the earth's surface—which is lucky for us. Explorer 1 began transmitting data, and the counts increased as expected. Then, at an altitude of 600 miles (970 km), they stopped completely. This happened at each orbit when Explorer 1 went over the critical height.

The situation was most peculiar, and at first Van Allen and his team did not know what to make of it. Eventually they found the answer. Above the critical height there were so many particles that Explorer 1's instruments simply could not cope with them, and had become over-saturated. In the magnetosphere—that is to say, the area round the earth in which our magnetic field is dominant—there are two zones of intense radiation. The inner belt, centered at about 1,900 miles above ground, is made up largely of energetic protons, due to interactions between cosmic ray particles and the upper atmosphere. The outer belt, centered at an altitude of around 12,000 miles (19,000 km), consists mainly of electrons captured from what is called the solar wind, a stream of particles being sent out from the sun. Spacecraft en route for the moon or planets pass through the Van Allen Belts quickly, although for an astronaut to stay in the zone for a protracted period of time would be decidedly unwise.

SATELLITES

By now so many satellites have been launched that to list them all would take many pages. They have been of various kinds and play so major a role today that it is strange to recall that 50 years ago they did not even exist.

Communications satellites are of special importance. One of the first to appreciate this was Arthur C. Clarke, known worldwide because of his science fiction writing and because of his association with the film *2001: A Space Odyssey*. As long ago as 1945 he wrote a short article in the magazine *Wireless World* in which he suggested setting up orbiting satellites to act as television relays. A satellite at a distance of 26,000 miles (42,000 km) will complete one circuit in 24 hours; if it is positioned over the equator, in a "geosynchronous" orbit, it will seem to stay motionless in the sky. Many such satellites now exist. Arthur has pointed out that if he had patented his scheme, he would be now be a multi-multi-millionaire!

It was in 1962 that the first TV pictures were beamed across the Atlantic from America to Britain, via the tiny satellite Telstar—which no doubt is still in orbit, although all track of it has long since been lost. Many people still fail to realize that all long-range television depends upon space methods; without satellites, you would be unable to sit in London and watch the Olympics in China.

Bodies in the sky send out radiations over the whole of the electromagnetic spectrum. Visible light makes up only a tiny part of the whole range, and most of the rest is blocked by layers in our upper atmosphere, although there are a few "windows" such as those at radio wavelengths. Without space techniques, the astronomer would be in the same plight as a musician trying to play a Strauss waltz on a piano which lacked all its notes except those of the middle octave.

The ultra-short gamma rays represent the most energetic forms of radiation. They were first detected from space in 1961 by the Explorer II satellite, and since then there have been several full-scale gamma ray space observatories. The results have been fascinating; for example we have recorded the very brief "gamma ray bursters," which last for only a few seconds, but which seem to be the most violent explosions that have occurred in the universe since the original Big Bang, 13.7 thousand million years ago. Then there are X-rays from space, which allow us to penetrate the dust around regions where new stars are being formed. We have also learned a great deal from ultra-violet studies; one satellite, the International Ultra-Violet Explorer (IUE) was launched in January 1978 and finally closed down in September 1996, not because it had broken down but as a cost-cutting exercise (to continue to monitor it for another 50 years would be almost as expensive as building a nuclear bomber). Infrared studies carried out from satellites have provided information that we cannot obtain in any other way.

All of these investigations have revolutionized astronomy, as well as having profound effects upon our everyday life. New discoveries are being made almost every month, but, sadly, we also have spy satellites and threats of space weapons—it is true that nuclear missiles could quite easily wipe out civilization. We can only hope that our own civilization will not self-destruct.

Opposite: Preparing Telstar, the first live communications satellite. It was roughly spherical, 35 ½ in. (88 cm) and weighed 170 lbs. (77kg). It was launched from Cape Canaveral on July 10, 1962, and at once relayed its first television picture; on July 23 it relayed the first live transatlantic signal. It went out of service in February 1963. Presumably it is still in orbit.

Above: A modern communications satellite, the Intelsat VI, is captured by the crew of the shuttle *Endeavour* in 1992. This was the most complex satellite rescue mission ever attempted, to repair its failing booster and return it to the correct orbit.

Top left: **1961**
August 13 – construction begins on the Berlin Wall, the separation barrier between West Berlin and East Germany (the German Democratic Republic), which closes the border between East and West Berlin for 28 years. The wall becomes an iconic symbol of the Cold War.

Bottom left: **1963**
November 4 – The Beatles appear on the Royal Variety Performance at the Prince of Wales Theatre, London. As their fame spreads, the frenzied adulation of the group, predominantly from teenage female fans, is dubbed "Beatlemania."

Top right: **1962**
The Cuban Missile Crisis begins on October 16, when U.S. reconnaissance data revealing Soviet near-range nuclear missile installations in Cuba are shown to U.S. President John F. Kennedy. The crisis ends 12 days later when Soviet premier Nikita Khrushchev announces that the installations will be dismantled and shipped back to the Soviet Union accompanied by a U.S. escort. The crisis is regarded as the moment when the Cold War comes closest to escalating into a nuclear war.

Bottom right: **1963**
On November 22, while driving through Dealey Plaza, Dallas, Texas, the 35th U.S. President John F. Kennedy, is shot dead by Lee Harvey Oswald.

1960–1969

Top left: **1965**
U.S. involvement in the Vietnam War escalates when 200,000 American combat troops are sent to South Vietnam to prevent the anti-Communist South Vietnamese government from collapsing. American policymakers believe that if the whole of Vietnam falls under a Communist government, Communism will spread throughout Southeast Asia—a belief known as the "domino theory."

Top right: **1968**
In Paris, France, student protests against university reform turn into riots and lead to the student occupation of the Sorbonne. The riots culminate in a general strike of students, teachers, and workers that almost brings down the government of President Charles de Gaulle.

Left: **1968**
August 21 – A Soviet-led invasion by the Warsaw Pact armies crushes the so-called Prague Spring reform movement.

Bottom: **1969**
Woodstock Music and Art Festival is held on a dairy farm in Bethel, New York, on August 16. The festival, which features many of the best-known musicians of the time—including Jimi Hendrix, Crosby, Stills and Nash, and Carlos Santana—exemplifies the counterculture of the 1960s.

Opposite: L. Gordon Cooper, Jr. enters his capsule, *Faith 7*, for the last of the United States' Project Mercury missions. His flight began on May 15, 1963 and lasted 34 hours, longer than all the other missions combined. He circled the earth 22 times, and became the first man to sleep in orbit.

02 MEN IN SPACE

In 1903 Orville Wright made the first manned flight in a heavier-than-air machine. Less than 60 years later Yuri Gagarin became the first man in space, and in the same decade Neil Armstrong and Buzz Aldrin stepped out on to the surface of the moon. It would have been possible for all four to have met face to face. They never did, but their lives overlapped—I know, or knew, them all.

The twentieth century was indeed a time of pioneering. Yet in its earliest years there were still many very eminent scientists who were convinced that human beings would never be able to leave the ground. Simon Newcomb, a leading American astronomer, wrote in 1904 that the only possible way would be to have an aircraft carried aloft by birds—which takes one back to Godwin's *Man in the Moon*; presumably Newcomb had read this, but did not know that the Wright brothers had already made their first flights.

THE DANGERS

Rocket developments made it clear that, from this point of view, manned spaceflight was possible, but there were many sceptics even after Sputnik 1 in 1957. There were several questions to be answered, and finding the answers was bound to be difficult without making a practical test. Zero gravity, radiation dangers, and meteoric bombardment all had to be taken into account.

As we have noted, cosmic ray primaries—that is to say, the heaviest particles—are broken up in the high atmosphere, which makes a very effective screen; only the shattered secondaries reach ground level, and these, unlike the primaries, are harmless—otherwise, no advanced life forms could have developed on Earth. Once above the atmospheric shield, an astronaut would be vulnerable, and nobody was sure how serious the risk would be. But short-wave radiations from space seemed to be an even greater threat. The sun is one such source, and for a long time astronomers had been keeping track of solar storms. There are violent outbreaks called flares that

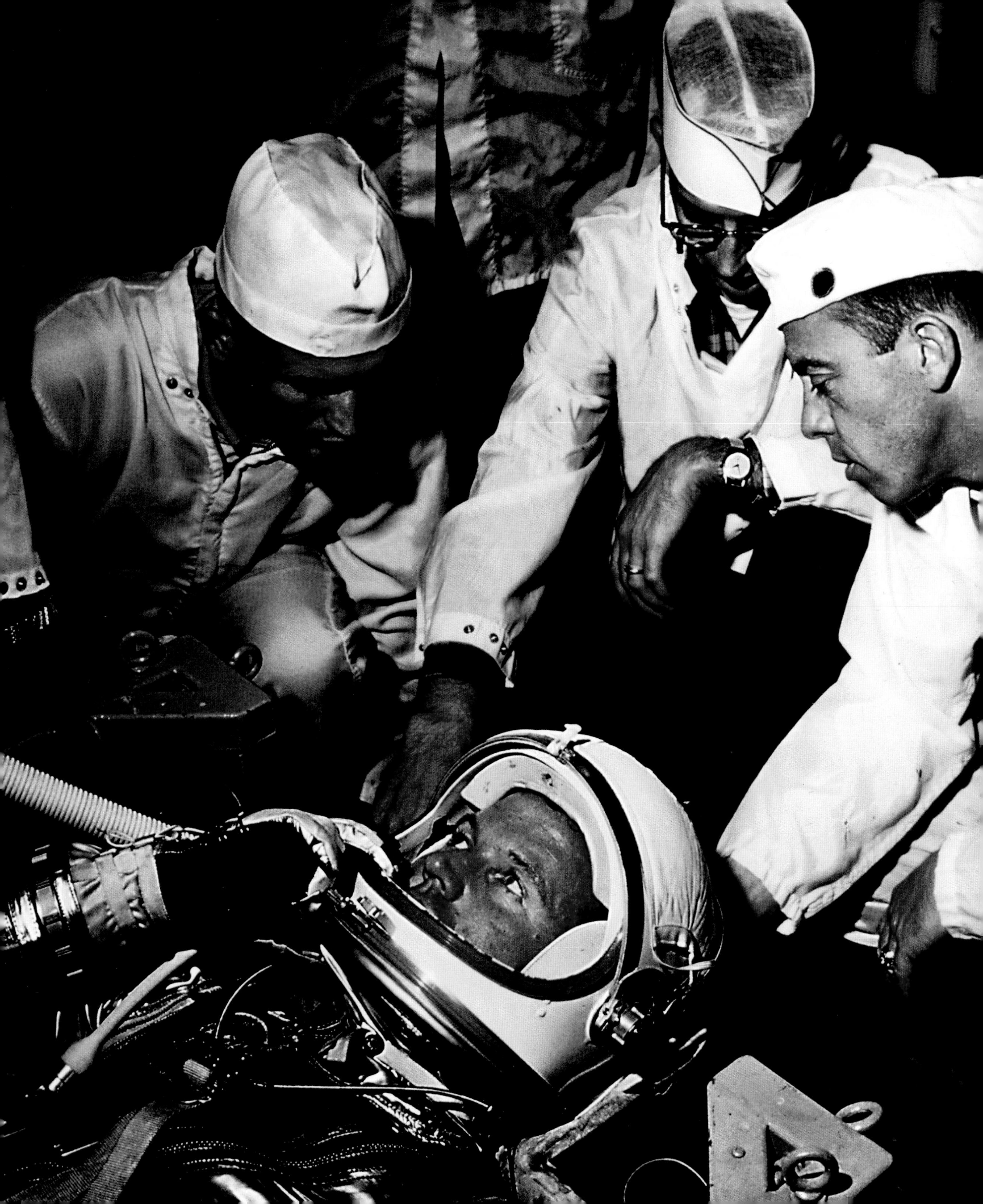

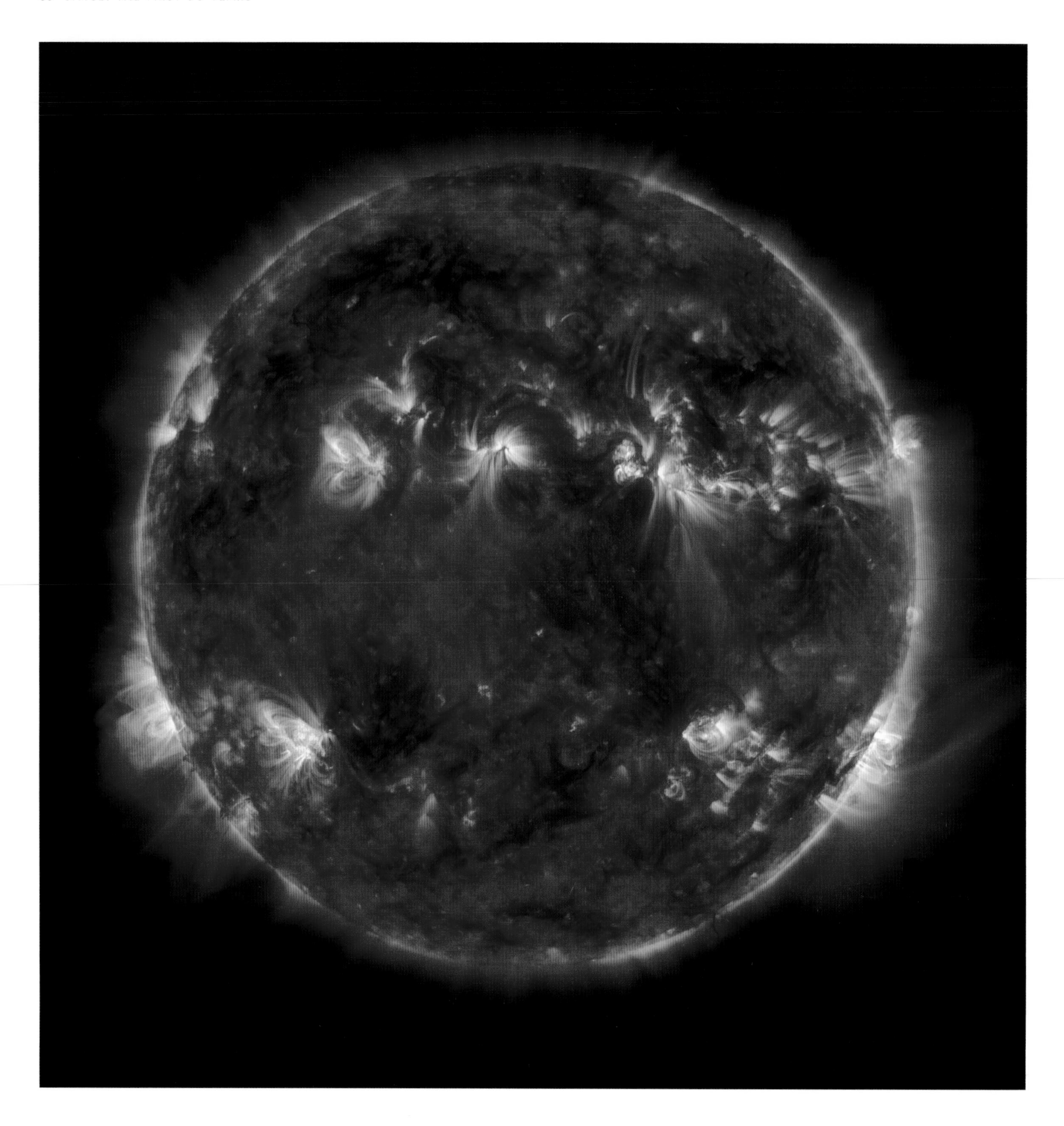

Above: Image of the sun taken from SOHO, the Solar and Heliospheric Observatory. SOHO was launched on December 2, 1995; it orbits in a stable position—the L1 (first Langrangian point), between the earth and the sun, just over 621,371 miles (1,000,000 km) from Earth. This image shows active regions on the disk, together with flares.

send out bursts of radiation as well as high-energy particles, which bombard the earth; they cause magnetic storms and interfere with radio communication as well as cascading downward to produce the lovely displays of aurorae. Since the particles are electrified, they make their way to the region of the magnetic poles, which is why aurorae are most commonly seen from high latitudes. From south England they are seldom striking, but they are much more frequent in Scotland, and a night in Antarctica of northern Norway would seem drab without them.

In a way the sun must be regarded as a variable star, with a definite short-term cycle of activity. At maximum there are many groups of sunspots, and frequent flares; at minimum there may be no major spot-groups for many consecutive days, and the sun is relatively "quiet." The average length of cycle is eleven years, and since the last maximum occurred in 2001 we may expect the next in 2012, but the cycle is not strictly regular, and moreover no one can be sure of events. A violent flare may occur at any time, and to an unprotected astronaut would be very dangerous indeed.

Next, there is the possible danger from small solid particles, of which there are plenty in the solar system. We can see them on any clear night, as shooting stars, which, as we have noted, are cometary debris. Though they are only of dust-grain size they are moving very quickly, and make quite formidable missiles. A well-made spacecraft can withstand them, but a larger object, a few centimeters across, would be much more of a problem. If a spacecraft were hit by a meteoroid the size of an office desk, for example, it would most certainly be so badly damaged that its crew would have no hope of survival. These bodies come from the asteroid belt, and are not associated with comets, and neither can they be seen from Earth; in 1957 we knew about them only because of meteorites, which ploughed through the atmosphere and landed more or less intact. For all we knew, space might be thickly populated by objects of a dangerous size.

There was also the problem of zero gravity, or weightlessness. Here, Jules Verne made a mistake in his novel. When his space travelers were sent up, at 7 miles per second (11.27 km per second), they were said to become weightless only when they reached the "neutral point" between the earth and the moon, where the pulls of lunar and terrestrial gravity cancelled each other out. In fact, Barbicane and his companions would have been weightless from the outset—assuming that they could have survived the shock of departure!

Consider a pencil resting on a book. It is pressing down, and with reference to the book. It is "heavy." Now drop the book. It falls down; until it hits the floor, it and the pencil are moving in the same direction at the same rate. The pencil is no longer pressing on the book, and with reference to that book it has become weightless. The situation would be the same if the book and pencil were moving upward at the same rate. In the same way, an astronaut inside a moving spacecraft will seem to be weightless. But how does the human body tolerate zero gravity?

Initial experiments were made using animals. The second artificial satellite, Sputnik 2, carried a dog, Laika, who was neither killed by radiation or by meteoric damage to the capsule, and despite zero gravity survived for several days in orbit before dying what we hope was a painless death. Quite a number of people (including me) voiced strong objections, because there was never any chance that Laika could make a safe return to Earth, but it cannot be denied that the experiment was very valuable scientifically, and paved the way for the first manned flight.

There was no shortage of volunteers, either in the West or the USSR. In Washington, the newly formed NASA set up Project Mercury, and selected their first astronauts, always remembered as the Original Seven. They were Alan Shepard, Scott Carpenter, John Glenn, Gordon Cooper, Walter Schirra, Virgil Grissom, and Donald Slayton. All flew in space, although Slayton had medical problems and had to wait until long after Project Mercury. Glenn was the first American to orbit the earth, and later, had the distinction of being the oldest astronaut when he made his second foray into space at the age of 77!

Training was rigorous by any standards. Each candidate had to have exceptional qualifications, both mental and physical; each was an experienced pilot of conventional aircraft, and each had to put up with extreme discomfort—such as being whirled dizzily around in a device termed a centrifuge. Each had to be scientifically competent, and to know every detail of the spacecraft to be used. Even so, it was difficult to simulate zero gravity for more than a very brief period, and there was always the dread of incapacitating space sickness.

THE FIRST MISSIONS

All seemed to be going according to plan, but a similar training in the East was in progress, and again the Russians were ahead. On April 12, 1961 Major Yuri Gagarin of the Soviet Air Force was launched in a capsule attached to an A1 rocket, and made a complete circuit of the Earth. His altitude above the ground ranged between 112 and 203 miles (180 and 327 km). The flight lasted for 1 hour 48 minutes. The capsule, *Vostok 1*, landed safely in the prearranged position; according to schedule Gagarin ejected from the capsule at 23,000 ft. (7,000 m) and ended his journey by parachute. It was not until several more tests that Russian cosmonauts actually remained inside their capsules until landing, though the Americans always did.

This was Gagarin's only space flight—sadly, he was killed in an ordinary aircraft crash seven years later—but its importance cannot be overemphasized. It convinced even the most stubborn sceptics that space travel was possible, and that the moon and other worlds were within reach. Terrestrial isolationism was over.

Less than a month after Gagarin, Alan Shepard in *Mercury 3* was launched from Cape Canaveral in a tiny, cramped capsule. He did not go round the earth; his flight was an up-and-down "hop" lasting for little over a quarter of an hour, but he reached an altitude of 116 miles (187 km) before splashing down in the sea 300 miles (500 km) from Cape Canaveral. The whole flight was broadcast, with a suitable running commentary, and was heard by millions of people all over the world, which was a welcome development (remember, this was during the Cold War). A second sub-orbital flight was made in July, with Virgil Grissom in the capsule; a month later came the Soviet Union's *Vostok 2*, in which Gherman Titov went round the world no less than 17 times, staying aloft for 25 hours. Incidentally, he was the first victim of space sickness, though it did not last for long and he recovered well before he was due to land. Then, at last, came the first American orbital flight, by Lieutenant Colonel John Glenn, in *Mercury 6*, better remembered as *Friendship 7*.

Several preliminary tests had been carried out, and on November 29, 1961, a chimpanzee named Enos had made two circuits of the earth, coming through unscathed though, according to reports, somewhat peevish. Like Shepard and Grissom, Glenn was to have his flight covered by television, radio, and the Press. There was nothing secretive about it. Had it gone wrong, there would have been no chance of a cover-up.

Glenn was launched on February 20, 1962. He was aloft for almost five hours, and found that weightlessness was not in the least uncomfortable; he even called it a "pleasant experience," and he soon adapted to it. His description of the scene is worth quoting:

> The horizon itself is a brilliant, brilliant blue and white. As the sun moves toward the horizon a black shadow of darkness moves across the earth until the whole surface, except for the bright band at the horizon, is dark. This band is extremely

Above: The "Original Seven" *Mercury* astronauts. From left to right: Alan Shepard, Virgil Grissom, John Glenn, Gordon Cooper, Scott Carpenter, Walter Schirra, and Donald Slayton. All made space flights, and Shepard went to the moon in *Apollo 14*.

Opposite top: Human centrifuge at the U.S. Navy Aviation Medical Acceleration Laboratory in Johnsville, Pennsylvania. It was first used for astronaut training during the *Mercury* program. A centrifuge puts the passenger into rotation round a fixed axis. The experience is very uncomfortable!

Opposite bottom: Effects on the passenger being whirled round in the centrifuge. Training of this sort is essential for an astronaut, who must be ready for very strong g-forces during a space flight—particularly during the launch, when the capsule is being sent upward at a velocity of several kilometers per second.

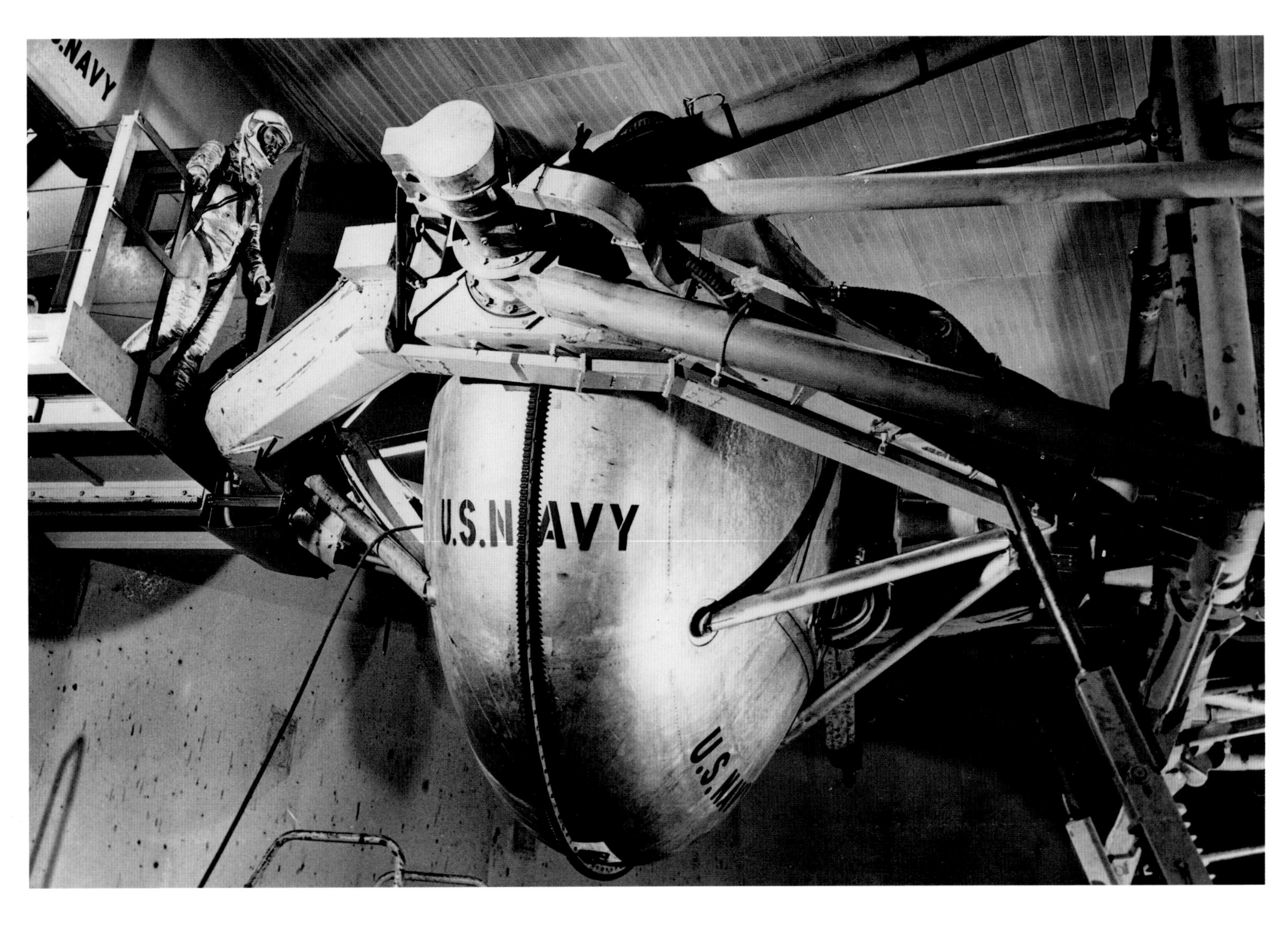

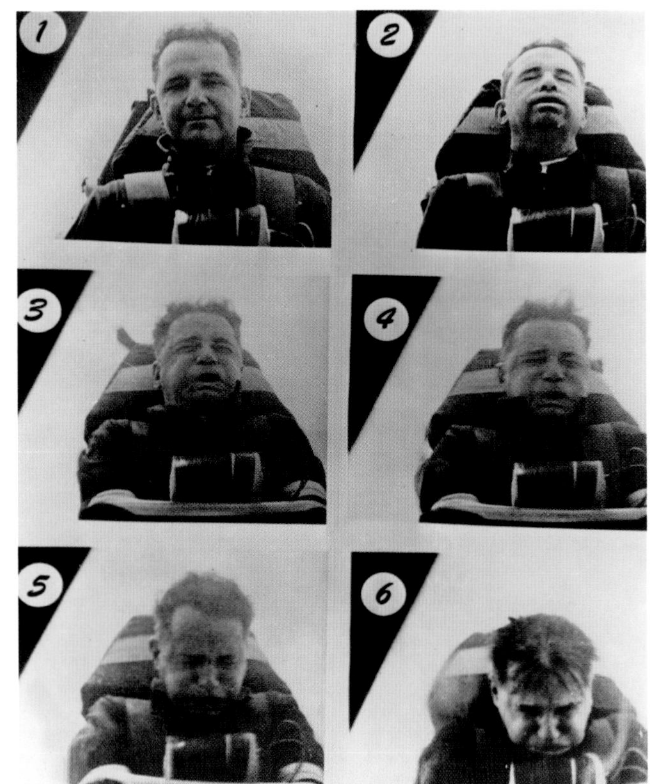

bright just as the sun sets, but as time passes the bottom layer becomes a bright orange and then fades into reds, then on into the darker colors, and finally off into the blues and blacks.

The colors seen when looking down at the ground were, he said, much the same as those seen flying in an aircraft at around 50,000 ft. (15,000 m). (In passing, there is a persistent rumour that the only artificial feature on Earth visible from deep space, or the moon, is the Great Wall of China. This is quite wrong. The Great Wall is none too easy to make out even from a jet aircraft on a commercial trip!)

All those early trips were very brief, but at least they showed us that some of the dire predictions put forward were very wide of the mark. For example, it had been suggested that as soon as an astronaut ventured beyond the atmosphere he would be killed by the effects of cosmic rays or harmful emissions from the sun, or perhaps his capsule would be fatally holed by bombardment from tiny, fast-moving solid particles; space sickness might strike him at once; his eyes would be affected by the harsh lighting conditions and so on. None of these fears appeared to be justified. When space missions became longer and longer, as they very soon did, the astronauts and cosmonauts were still unaffected, though after return to Earth it took a while to readapt to being "heavy" again. On the other hand, this did not rule out the possibility of longer-term damage, and there is still considerable uncertainty about this. A journey to Mars, taking months, is very different from orbiting the earth at a distance of a few miles. However, at the start of the space age problems of this kind lay far in the future, and certainly the outlook was encouraging.

Events moved quickly during the 1960s—so quickly, in fact, that there were grandiose ideas of lunar bases and Martian expeditions before the year 2000. To give a full account of those heady years would take many pages, but it is remarkable that men reached the moon less than a decade after Gagarin's first flight. Here, then, are some of the highlights—and the lowlights.

The cramped, one-passenger capsules of the *Mercury* and *Vostok* period were succeeded by larger crafts able to carry several people. Weightlessness did not seem to be either harmful or uncomfortable, but it did pose problems, and the astronauts had to get used to them. Eating and drinking were problematic. A weightless liquid will not pour and at first the space menus were unpalatable, although as time went by things were better, and eventually meals in space became similar to those at home. Then, a delicate question—how does one "go to the bathroom" in space? This was solved fairly easily; waste products were collected in bags by suction, to be stored and disposed of after return.

Two events marked 1963. For the first time, two manned spacecraft were in orbit at the same time: Andriyan Nikolayev in one, Pavel Popovich in the other. Valentina Tereshkova, also from the USSR, became the first woman in space. Two years later came the initial "spacewalk" by Alexei Leonov. At this stage the Russians seemed to be first in all major developments, but the Americans were not far behind, and, in fact, were well ahead in preparations for sending men to the moon.

On March 18, 1965, Pavel Belyayev and Alexei Leonov went up in *Voskhod 2*; Belyayev was the commander, but it was Leonov who took the limelight. During the mission he put on a spacesuit and went outside the *Voskod* to make the first space walk. He did not go far, and he did not stay out for long, but he certainly made history.

Of course, he carried a safety cord, but there was no tendency to drift away, because he and the capsule were moving in the same orbit at the same speed; the situation may be likened to that of two ants crawling on the rim of a rotating bicycle wheel. Later he wrote: "Nothing will ever compare to the exhilaration I felt. No matter how much time has passed, I can still remember clearly my conflicting emotions. I felt almost insignificant, compared to the immensity of the universe." But his foray nearly ended in disaster. He was outside the spacecraft for 12 minutes, and when he prepared to reenter he found that his spacesuit had inflated in the vacuum to the extent that he could not go through the airlock. As he admitted later, it was a terrifying moment. He had to risk reducing the suit pressure to a dangerously low level, and even then he was only just able to scramble inside.

Leonov, one of the most popular and most articulate of the cosmonauts, has had a distinguished career. If the Soviet rocket program had gone according to schedule he would have been the first Russian on the moon, and perhaps the first man of any nationality. Even now, in his seventies, he remains actively involved in space planning. He is, incidentally, an extremely good "space artist."

TRAGEDIES

In the opening period of the new age there had been no fatal accidents. When tragedy struck, on January 27, 1967, it was not in space, but a mere two hundred feet (61 m) above the ground. The first flights of the *Apollo* moon program were in an advanced state of planning, with a crew consisting of Virgil Grissom, Edward White, and Roger Chaffee; Grissom was one of the Original Seven, and White had been the first American spacewalker, a few weeks after Leonov. During a routine rehearsal the three were in their takeoff positions when there was a sudden outbreak of fire—too quick and too violent for any rescue operation to be carried out. The atmosphere inside the cabin was pure oxygen, and the astronauts had no chance. They had been well aware of the risks they were running, but it was doubly tragic that they died during a rehearsal which ought to have been completely safe.

Opposite: Yuri Gagarin, the first man to go into space. He made only one orbit, but his flight was immensely significant: he was not spacesick, he was not killed by radiation, his vehicle was not battered by meteoroids, and he had no problem with weightlessness.

Above: Launch of *Vostok 1*, sending Yuri Gagarin around the earth. This proved that spaceflight is possible, and that other worlds are within our reach. It also showed that at that time the USSR was well ahead of the United States in what was generally termed the "space race."

Sadly, another tragedy soon followed, this time actually in space. When Vladimir Komarov of the USSR was launched in the latest Russian spacecraft *Soyuz 1*, there seemed to be no cause for apprehension; Komarov had already made one flight, and was known for his exceptional skill. But things went wrong almost at once. When Komarov prepared to land, the equipment failed; the parachutes did not deploy as they should have done, and *Soyuz 1* crashed down. The pilot could not be faulted; the trouble lay with the spacecraft itself.

Space is a dangerous environment. Yet it is worth noting that fewer people have been killed in the development of astronautics than died during the early story of aeronautics; so far there have been over 400 space travelers, and fewer than twenty casualties. Comparing this with the daily carnage on our roads makes space flight look comfortingly safe!

Over the years there have been persistent rumors that the Russians have had many more casualties than they care to admit. It has been claimed that Gagarin was not the first man to go into space, but merely the first to survive. It is true that in the Soviet era the Russians were disinclined to let the outside world know what they were doing, but the stories of lost astronauts are based upon evidence which is very slender indeed, and on balance it seems most unlikely that Gagarin had any predecessors, successful or otherwise.

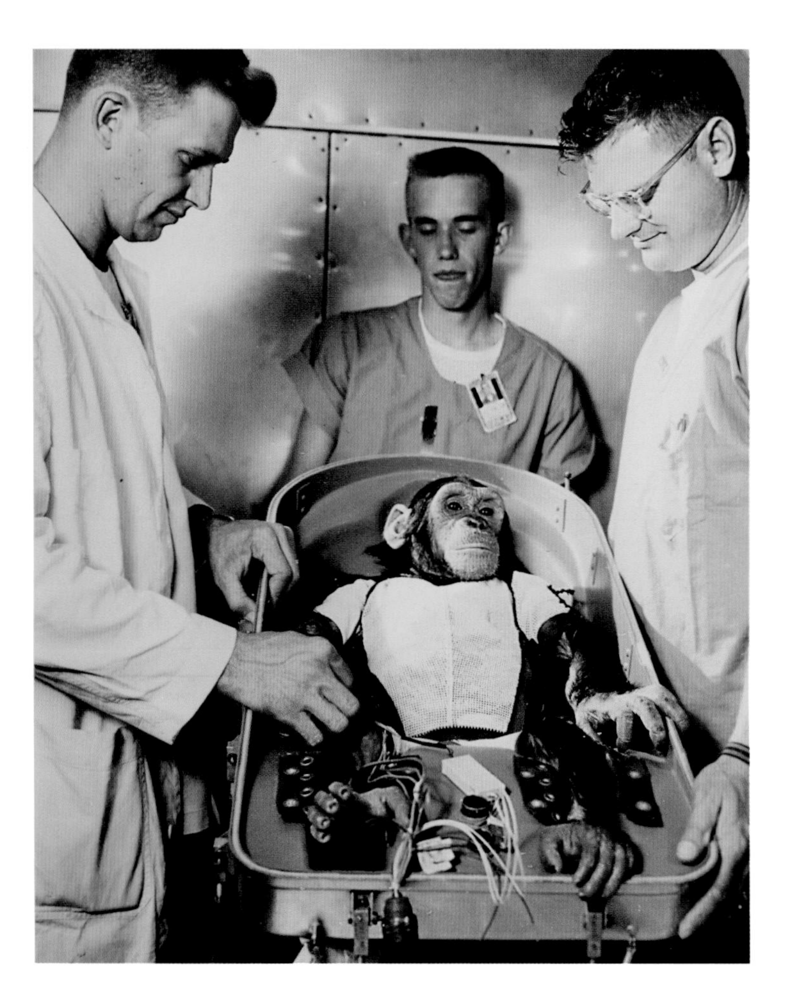

SCIENTIFIC RESEARCH

In one respect things are better today than they have ever been before, because space research has become genuinely international. Information is shared, and although the United States and Russia have been responsible for over 99 percent of launchings other nations have started to join in, notably Japan, France, and most recently, China. Many nationalities have been represented in space, some of whom—notably the British-born, Cambridge-educated Michael Foale—have played extremely important roles. The problem of military space research has not disappeared, but it has been reduced. If humanity really wants to wipe itself out, it can easily do so by using weapons of mass destruction already stockpiled by the Pentagon, the Kremlin, and no doubt elsewhere. We can only hope for the best.

Following the first spacewalks it was evident that astronauts could cope well with conditions in space. Moreover, the spacecraft themselves could be maneuvered, and could dock with each other; then came the *Apollo* missions. But before discussing these, let us go forward to the early 1990s, to the story of the rescue of the Hubble Space Telescope.

The concept of a telescope orbiting the earth above the atmosphere originated with the American astronomer Lyman Spitzer, as long ago as 1946. He pointed out that seeing conditions would be perfect all the time, and there would be no trouble from everyday

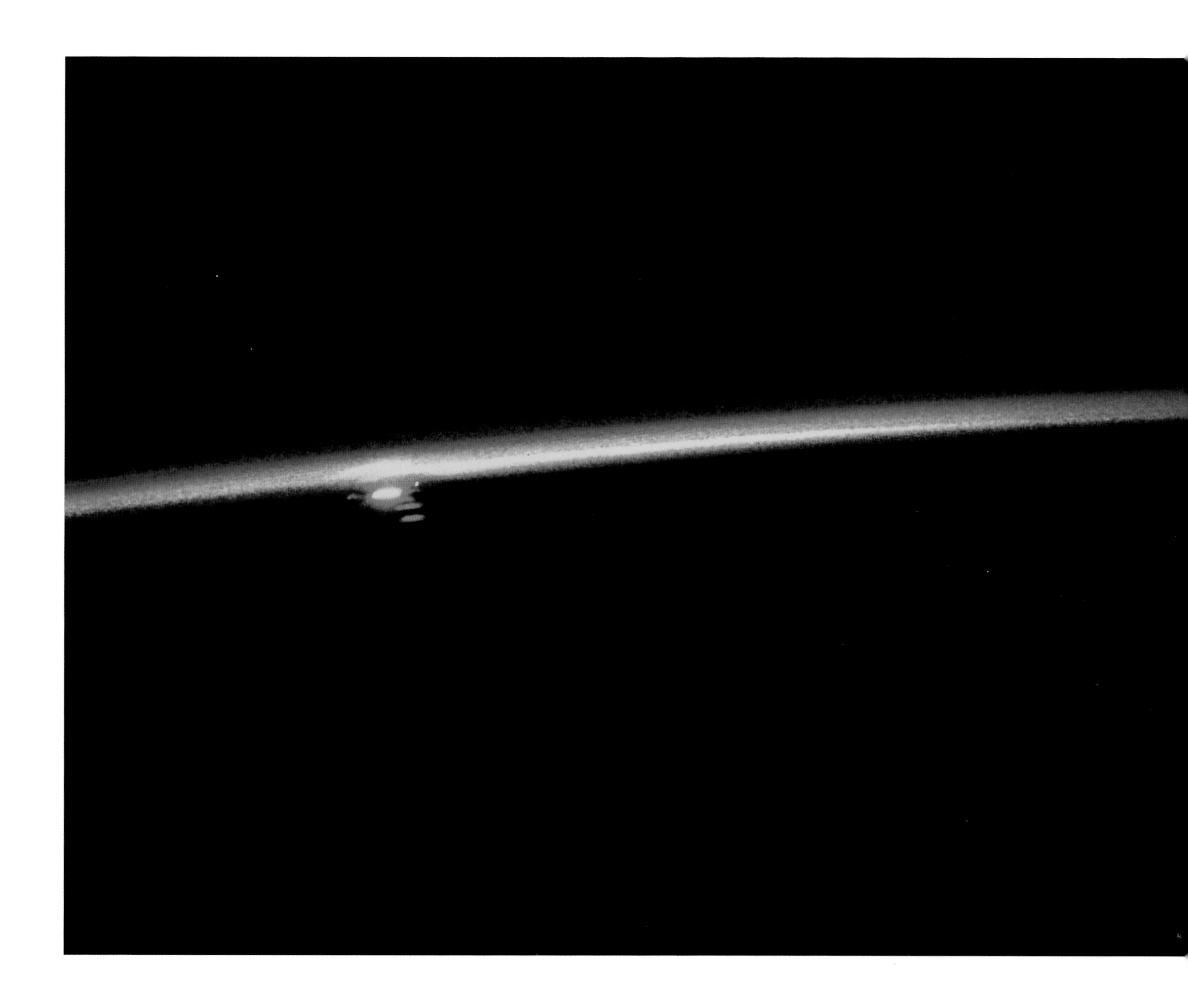

Opposite bottom: Enos, the chimpanzee astronaut, being prepared for his space flight on November 29, 1961. He completed two orbits and landed safely. Enos was put into honorable retirement. Chimpanzees were favored for experiments because they are so closely related to humans.

Opposite top: John Glenn, the first American in orbit, entering his capsule, *Friendship 7*, on February 20, 1962. He completed three orbits and was aloft for 4 h 55 m 23 s, reaching a maximum altitude of 162 miles (261 km). His flight was remarkably trouble-free. Glen made another trip into space at the age of 77, to test the reactions of older people. He returned unharmed.

Above: Sunrise, as seen by an astronaut in orbit. Note the clearly defined bands of color as the rays of the rising sun pass through the atmosphere. This is one of the best examples of "twilight layering," taken by the *Skylab 3* astronauts from an altitude of 270 miles (435 km).

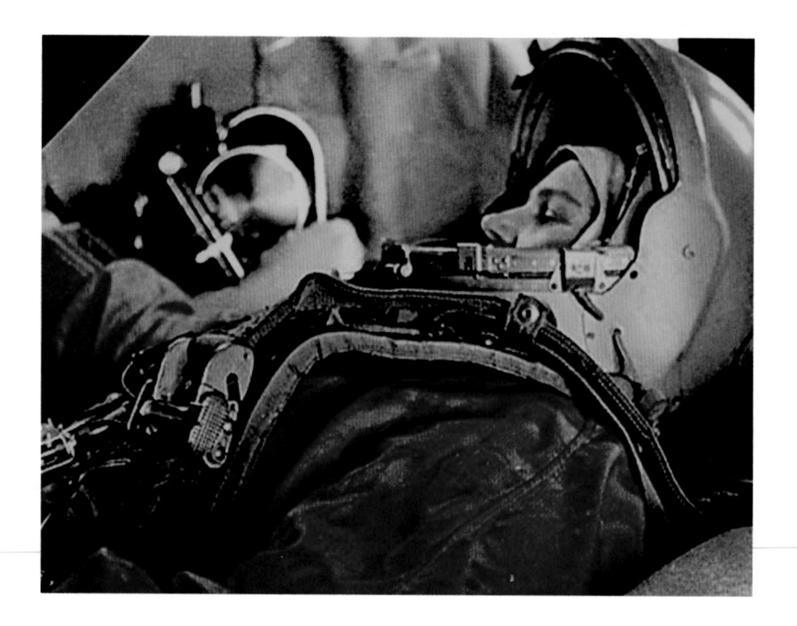

annoyances such as light pollution. It was not until the 1970s that such a project became practical, and even then there were constant delays, financial and otherwise. The telescope, named in honor of Edwin Hubble, was finally launched from Cape Canaveral on April 24, 1990. It was, in itself, a conventional reflector, with a 94-inch (239-cm) mirror. It used a well-tested optical system known as the Ritchey-Chrétien. The focal length was 189 ft. (58 m). It was much smaller than some ground-based telescopes, but moving in a circular orbit at a height of 360 miles (579 km) above ground level it was expected to outperform any telescope previously built. The actual launch was faultless; Hubble soared into the correct orbit, and the astronomers and the space planners purred with satisfaction. All seemed to be well.

Alas, all was very far from well, as was painfully clear as soon as the first star images were transmitted. There was a serious fault—what is known as a spherical aberration. Instead of appearing as a sharp point, a star showed up as a fuzzy halo about 4 arc seconds across. This may not seem much—one arc second, or 1/3600 of a degree, corresponds to the distance between the two headlights of a car as seen from a range of 300 miles (483 km)—but it was enough to wreck most of the scheduled observational work. The cause was soon found: human error. The mirror had been wrongly figured, and was too flat near the edge by about 1/50th the width of a human hair. The culprits were never identified, thanks to a cover-up worthy of any modern politician.

What followed was greatly to NASA's credit. On December 2, 1993, the space shuttle *Endeavour* took off carrying a crew charged with carrying out the most difficult and dangerous mission ever attempted. The telescope had to be captured and brought into the bay of the shuttle, where it was fitted with a new optical system to compensate for the defect in the mirror. Altogether five spacewalks had to be made, and in some of these the astronauts were untethered to the spacecraft. The operation was a triumphant success; Hubble was put back into orbit, and was not only cured, but functioned better than had been expected at the outset. Since then there have been other servicing missions, but none with the drama and tension of the first. At the time of writing (2007) the telescope is still in orbit, and still sending back magnificent pictures and invaluable data.

People from Earth had reached outer space. And they had every intention of staying.

Above left: Valentina Tereshkova, the Soviet cosmonaut who became the first woman in space; in June 1963 she piloted *Vostok 6*. She made 48 orbits, and stayed aloft for almost three days. Two days earlier *Vostok 5* had been launched, piloted by Valeri Bykovsky; the two cosmonauts were orbiting at the same time, and at one stage were only 3 miles (5 km) apart.

Above right: Cosmonaut Alexei Leonov was the first person to conduct a spacewalk. On March 18, 1965, on the Soviet Union's *Voshkhod 2* mission. Leonov disembarked from the two-man craft, and spent about 20 minutes floating in space tethered to the *Voshkhod*.

Opposite: On January 27, 1967 tragedy struck the *Apollo* program. Astronauts Virgil Grissom, Edward White, and Roger Chaffee were taking part in a rehearsal in a command module on the launch pad when a sudden flash caused a fire, and there was no chance to reach the astronauts. The tragedy led to drastic modifications to *Apollo* modules.

Opposite: Computer simulation of the completed *International Space Station* (*ISS*). It is immensely elaborate—and it really is international, something which would have been unthinkable in the days of the Cold War.

03 STATIONS IN SPACE

Tune in your television set, and you will be able to find a weather forecast either for your own area or for any part of the world. It may not be fully accurate, but it is not likely to be very wide of the mark. Thanks to satellites, we know far more today about the behavior of the atmosphere than we did a few years ago. Geostationary satellites orbiting above the equator at 22,300 miles (35,800 km) can monitor the whole of the world, and keep track of entire weather systems; they can also give advance warning of dangerous storms developing out at sea, giving people in the vulnerable areas time to move to a safer place. Many lives have been saved in this way. Forest and bush fires can be spotted as soon as they break out; areas of diseased crops can be identified, and so on. The first weather satellite, Vanguard 2, was sent up in 1959. Observations of the earth from space have revolutionized the whole science of meteorology, and have also been invaluable to geologists, geophysicists, oceanographers—in fact, to scientists of almost all disciplines.

Similarly, unmanned satellites have been of the utmost value to astronomers. But what about manned satellites? Are they practical? Are they really needed? Is there any justification for building a space station? Arguments about this have raged for many years, and are likely to continue for a long time yet.

In post-war times, the first really detailed plans were put forward by Wernher von Braun. His proposed station took the form of a wheel, with the control center in the hub and the astronauts housed in the rim; rotating the wheel would simulate gravity in the form of centrifugal force. This would mean that the crew members would not have to remain "weightless" for long periods. The concept took hold of the popular imagination, if only because a von Braun wheel would be so graceful! But at that time, even Sputnik 1 lay in the future, and when space stations were eventually built they were much less aesthetically appealing.

USA
PK

There was one obvious point to be taken into account. If a space station is to be occupied, there must be up and down transportation for the astronauts, and a reusable vehicle would have to be designed (after all, it would be rather uneconomical, for example, to build a completely new train for every journey between two cities). The Space Shuttle program was started by NASA in the late 1960s and has dominated all manned operations ever since. It has cost much more than expected, and has been less reliable; there have been two major disasters, and innumerable delays and complications, but the Space Station program would have been impossible without it or some equivalent. The Russians were working upon shuttle-type vehicles at the same time. Collaboration between the Soviet and the American teams would have benefited both—but this was the era of the Cold War.

A NASA shuttle was made up of a reusable orbiter, the expendable external tank, and two reusable solid-fuel rocket boosters. Takeoff was vertical; the tank and the boosters were jettisoned during the ascent, after which the orbiter was left on its own to dock with the space station, firing its own engine to break free at the appointed time and reenter the atmosphere, landing unpowered in the manner of a glider. The shuttles were given attractive names—*Columbia, Challenger, Atlantis, Discovery, Endeavour*—but although many successful flights were made, *Challenger* was destroyed during launch in 1986, and *Columbia* broke up during reentry in 2003. In neither case was there the slightest hope of rescuing the crews. It may well be that the Russian shuttle, *Buran* ("Snowstorm") was better than NASA's, but it made only one flight—unmanned—in November 1988, after which the impending disintegration of the Soviet Union killed off the whole project.

SALYUT AND SKYLAB

The Russian *Salyut 1*, launched on April 19, 1971, was the first space station, sent on its way by a proton rocket. Six more *Salyuts* followed. A *Salyut* ("firework") was simple and could accommodate a crew of three. The program was partly scientific, partly military, and had very mixed fortunes. The first *Salyut* was occupied for 23 days, and all went well—until the cosmonauts boarded their ferry rocket *Soyuz 11*, for the return to Earth. The *Soyuz* landed intact but a structural fault allowed the inside air-pressure to fail and the crew members were dead on arrival.

Top right: Space station design by Wernher von Braun, 1952. The crew quarters were in the rim of the wheel; rotating would produce "artificial gravity." Nothing could be less like the *International Space Station* of today.

Bottom right: January 5, 1972. U.S. President Richard Nixon and NASA administrator James C. Fletcher announce that the Space Shuttle program has received final approval (San Clemente, California).

Opposite: The Space Shuttle has made building space stations more straightforward. Here, *Atlantis* delivers the *Destiny* Laboratory Module for the *International Space Station* in 2001.

Salyut 2 (April 1973) was a total failure, and was never manned, but its successors fared better, although the Soviet authorities gave out little information about them. The last *Salyut*, No. 7, was launched on April 19, 1982. By the time it reentered the atmosphere, on February 7, 1991, the *Mir* space station was in full order.

The first American space station was *Skylab*, sent up from Canaveral on May 14, 1973. The launcher was a huge Saturn-V rocket; a vehicle of the type used in the *Apollo* program. There were immediate problems. An important shield was torn free during ascent by atmospheric drag. *Skylab* settled into its planned near-circular orbit, 270 miles (430 km) up, but in a decidedly battered condition. The first crew followed separately, in a shuttle, and docked; they had to carry out emergency repairs, which included fitting what can best be described as a sunshade. Under the circumstances, it is remarkable that *Skylab* lasted as long as it did and accomplished as much as it did.

Skylab stayed aloft for over six years, completing 34,981 orbits of the earth, but its useful life was limited to the period between its launch date and February 1974, when the last of its three-man crews left it. The now unmanned vehicle eventually fell back into the atmosphere. It broke up during its final plunge, and scattered debris over Western Australia, fortunately without causing any damage. Actually it came down sooner than NASA had predicted as the density of the upper air was greater than they had expected, which, of course, increased the drag.

During its active life the station was occupied for a grand total of 171 days. Two of the astronauts who went to it, Pete Conrad and Alan Bean, had already been to the moon, with *Apollo 12*, so to them *Skylab* must have seemed almost parochial. A great deal of scientific work was carried out, including some medical experiments, and 10 space walks were made. Observations of the sun were particularly valuable, and what are now called "coronal holes" were discovered; these are very low-density areas in the sun's corona through which the solar wind particles can escape into space.

MIR AND EAST–WEST COOPERATION

While *Skylab* and the *Salyuts* were designed as temporary bases, *Mir* (a name which in Russian can mean either "world" or "peace") was much more substantial, and the intention was to man it for several years. It was launched on February 19, 1986, and did not come to the end of its career until March 23, 2001. In all, it stayed up for 5,511 days and was occupied for 4,594 of them. It completed just over 89,000 orbits at a height of 240 miles (430 km) above the earth. Its orbital period was 89 minutes, and since it was easily visible with the naked eye it became a familiar feature of the night sky.

There are several important points to bear in mind. First, *Mir* was a triumphant success—despite some unfavorable reports in the popular press. Its scheduled lifetime was seven years, and during this whole period it functioned excellently. Later, it had a great many problems,

Below: *Skylab* was America's first, and so far only, space station, and was launched in 1973. This view shows the station against a black sky, with its solar panels deployed (the one on right failed to open properly and broke off).

but all these occurred when the station was well past its "sell-by" date. It remained in orbit for much longer than originally intended because of the changing political scene at home. In 1986 Mikhail Gorbachev was in power in the Soviet Union; the Cold War was still ongoing, admittedly not with the intensity of earlier years, and space collaboration was very limited. Certainly there was no prospect of American astronauts visiting Soviet stations, or vice versa; there had been one link-up, but the two great powers remained very wary of each other. All this changed with the disintegration of the USSR, and joint missions were agreed. A docking module was attached to *Mir* so that American shuttles could link up with it. The Russians had relied upon their much less elaborate proton servicing rockets.

The space station was built by connecting several *Mir* modules, each launched separately by protons. First came the Core module (February 19, 1986), which provided the living quarters for the occupants. In 1987 came Kvant 1 (astronomical research), in 1989 Kvant *2* (improved life support systems), in 1990, Kristall (material processing, with new scientific laboratories) and then, in 1995, Spektr (general experiments involving the American program). The docking module was taken up by the U.S. Shuttle *Atlantic* on November 12, 1995, and finally came Priorda (remote sensing module) on April 23, 1996. *Mir* was then complete. It was a strange looking structure, very different from Wernher von Braun's picturesque wheel, but it was undeniably efficient.

First, and perhaps foremost, *Mir* showed that humans can remain weightless for long periods of time. On *Mir*, the Russian cosmonaut Dr Valeri Polyak stayed on the station for 14 months (to be precise 437 days), and on return did not take long to readapt to conditions of normal gravity. This is not to say that long-term effects can be ruled out, but at least the problem is less immediately pressing than was once forecasted. Needless to say, all possible precautions are taken, and "exercise machines" are important items of a space station's equipment.

It was said that *Mir*'s interior looked like a general store, crowded with scientific instruments, cables, and hoses, as well as everyday items. Normally there were three occupants, though on occasion it was home to as many as six for a few weeks. Quite apart from Russia's cosmonauts and NASA's astronauts, there were many visitors from other nations, some of whom were genuine researchers while others were really tourists—though the first official space tourists did not come on the scene until after *Mir*'s time. Between launch and August 1999 the station was continuously manned apart from two short breaks.

Scientific results were of the greatest value. Medical research was obviously a priority; so were biological and chemical experiments. Conditions for astronomical observations were ideal; the earth, far below, could be continuously modified—the scope was indeed wide, and *Mir* did everything that had been expected of it. Only after the space station had exceeded its scheduled lifetime did things start to go badly wrong. The situation was not helped by the fact that the new Russia, unlike the old USSR, did not have virtually unlimited funds to spend on space research.

Fire broke out in February 1997, and there was even a chance that *Mir* would be abandoned, but the crew managed to keep the blaze under control, and extinguish it before it could spread. But pieces of equipment began to fail; *Mir* was showing its age. There was a crisis of a different sort on June 25, 1997, when an unmanned Progress supply rocket collided with the Spektr module, puncturing it and causing severe damage. One astronaut commented that carrying out repairs to *Mir* was rather like working on an ancient second-hand car, and the lives of the occupants became less and less comfortable. Yet the scientific work went on for as long as possible and it is true that there was no shortage of volunteers. One man who played an important part was the British astronaut Michael Foale, who undertook a particularly hazardous space walk to make essential repairs to the station's power supply.

By 1999 it became clear that action would have to be taken. The ideal solution would have been to boot *Mir* into a much higher orbit, well above the atmosphere, and simply leave it there until the development of technology adequate to reactivate it, but the money was not forthcoming, and, of course, the maneuver would have been far from easy. With regret, the decision was taken to bring it down. On August 27, 1999, the last station commander, Sergei Avdeyev "turned out the lights for the last time" and apart from one routine visit in April 2000 there were no more occupants. *Mir*'s own power was used to send it on its final, suicidal plunge into the atmosphere where it broke up. Its fragments fell into the sea between Chile and New Zealand, and the story of the first major space station was over. However, by then the construction of the *International Space Station* had begun. Without *Mir*, this certainly would not have been possible. Zarya, the first section of the *ISS*, was sent up from the Baikonur cosmodrome in Kazakhstan on November 20, 1998.

THE INTERNATIONAL SPACE STATION

It is refreshing to find that the *International Space Station* really lives up to its name. It began as a joint project of five space agencies—those of Europe, the United States, Russia, Japan, and Canada, with participation also from Brazil. Up until now the arrangement has worked well, apart from the fact that (predictably) the construction is years behind schedule, and the cost is far greater than expected. (Remember though, that the money spent on the invasion of Iraq would have financed the entire space program well into the twenty-second century!)

The *ISS* moves round the earth at an altitude of 220 miles (350 km). This varies slightly partly because the orbit is not perfectly circular and

partly because of changes in atmospheric drag; even at this height there is still a trace of air left. The period is only nine and a half minutes, so the *ISS* goes round the world over fifteen times per day. When well-placed, it is a prominent object visible to the naked eye. It will be four times the size of *Mir* when completed. The first crew arrived there in November 2000, and since then there have always been at least two occupants. The names of the members of the original crew are worth remembering: two Russians (Sergei Krikalev and Yuri Gigzenko) and one American (William Shepherd). Krikalev was a space veteran. When he made his first flight, the USSR was still intact.

Building the *ISS* involves over forty separate flights. The main problem to date has been the unreliability of the shuttle program, plus the fact that the Russians are always short of money. On the plus side, there have been no really important problems with the station itself, and there is no reason to doubt that the entire station will be complete by 2010. It will then look rather like an enlarged and modified version of *Mir*, with 10 main pressurized modules with smaller sections attached to them. There will always be two *Soyuz* vehicles to act as lifeboats in the event of a sudden emergency—something that cannot be ruled out. In particular, no spacecraft has yet been disabled by collision with a meteoroid, but we have to accept that sooner or later this is bound to happen.

The science programs will be extensions of those carried out on *Mir*, but there have been great developments over the last few years. The techniques of 2006 are very different from those of 1996, and by the time that *ISS* is complete these techniques will be different again, so that research programs will be amended accordingly. There have already been more visitors than the total number of people who went to *Mir* throughout its whole career—and there have been official space tourists of whom the American businessman Dennis Tito was the first. He at least has no financial problems—which is just as well, because his return ticket to *ISS* cost the trifling sum of U.S. $20,000,000! There has

Opposite: Inside the *Mir* space station. One has to admit that it is cluttered like a corner shop; there was not much room for crew members and their equipment! Much was heard about the problems of *Mir*, but most of these occurred after the station had been in orbit for much longer than originally planned and a vast amount of important reseach was carried out during its lifespan.

Below: *Mir* orbits the earth. Despite offers for private funding from captains of industry and television stations, *Mir* was abandoned in August 1999. Late into the night of March 22 a series of engine burns began to nudge it down into the earth's atmosphere, where it broke up and was incinerated.

also been a space wedding, when cosmonaut Yuri Malenchenko married Ekaterina Dmitriev. Admittedly, the bride was not on *ISS* at the time; she took her vows from her church in Texas.

International though it was, the *ISS* had actually been built by the Russians and the Americans, who had dominated space research since the very beginning. Yet this state of affairs might not continue, because well before the end of the century the Chinese were again starting to take a lively interest. China had had plans for a space program for years, but during Chairman Mao's cultural revolution all this was put on hold. When things reverted to normal, it did not take the Chinese long to start catching up with the rest of the world. The Long March rockets worked well, and the first Chinese artificial satellite, Dong Feng Hong ("The East is Red") went up in April 1970. Others followed, and in 1992 the Chinese leaders officially authorized a manned space program, beginning with the development of the *Shenzhou* vehicles. On October 15, 2003, Colonel Yang Li-Wei completed fourteen orbits of the earth in *Shenzhou-5*, launched from the Jinquan Satellite center. Then, on October 12, 2005, *Shenzhou-6* carried two "taikonauts," Fei Junlong and Nie Haisheng, on a trip that lasted for five days. (Incidentally, this was the first time that Chinese spacemen had been provided with toilet facilities!) *Shenzhou-6* landed at Hongger Township, Inner Mongolia, after a trouble-free flight.

The authorities were suitably enthusiastic, and the chief Chinese rocket designer, Wang Yongzhi, announced that a permanently manned space station would be set up in the foreseeable future—certainly before 2010. There is no reason to suppose that this forecast is overly-optimistic. The Chinese have not, as yet, given any indication that they are ready to cooperate with the West, but they have, at least, joined the International Astronomical Union, which is encouraging.

Yet, despite all these developments, the whole program of manned space research has been strongly criticized by some very eminent scientists. The *ISS* has even been described as the white elephant of astronautics—on a par with the Millennium Dome in London. It has been claimed that the money spent on manned projects contributes very little to science, and is holding up research which is more productive. Arguments between supporters of these two diametrically-opposed schools of thought have become decidedly heated, and, as is often the case, things are by no means clear-cut.

Firstly, what can robots do that men cannot? Certainly they can be subjected to risks that would be unacceptable for a human being: to lose a rocket or space probe is unfortunate, but no more. Moreover, a rocket can venture where a piloted vehicle cannot. There would be few volunteers for a mission to Venus, with its crushing atmosphere, its intolerable heat, and its searing acid, whereas Russia's *Veneras* have landed on the surface and transmitted uncomplainingly—though admittedly they did not remain active for long before falling silent. A robot can send back data from vast distances; the *Pioneer* and *Voyager* probes have remained in contact when they have moved out far beyond the orbit of Neptune. And an automatic vehicle can be sent on a suicide mission, as with the "entry" of the Galileo probe to Jupiter.

Cost is all-important, and the opponents of manned research are quick to point out that a robot is much cheaper than a vehicle carrying a passenger. This is true enough, but it is worth noting that in NASA's most expensive year, when the *Apollo* program was in full swing, the Americans spent the same amount on space research as they did on military intelligence. One does not need to be an Einstein to see where financial savings could justifiably be made. On the other hand, it would be wrong to concentrate upon manned research and cut right back on the use of robots.

Now look at the other side of the coin: where are robots a disadvantage? They are unable to make on-the-spot decisions, and they cannot interpret information to the same extent as a man. A trained geologist, Harrison Schmitt, went to the moon with *Apollo 17*, and his specialist knowledge was of the utmost value. (This is not to decry the efforts of the other *Apollo* astronauts, all of whom were highly competent in all the relevant branches of science.) Remember, a machine is only as good as its maker.

If we want to make real progress, it is surely reasonable to accept that we need both manned and unmanned space programs; it is not a question of one or the other; the two are not mutually exclusive.

Meanwhile, construction of the *International Space Station* is fully underway. Go out and look, at a suitable moment, and you will see it passing far above you; on board, the astronauts will see the earth below. Less than a century ago this would have belonged solely to the realm of science fiction, but today it is accepted as part of modern life. We have come a long way since Sputnik 1 lifted off in October 1957.

Opposite: Astronaut James S. Voss handles the newly-delivered main boom of a Russian crane as part of a 6 h 44 m program of construction and repairs to the *International Space Station*.

Above: In the early 1990s, the end of the space race enabled Europe, Russia, Japan, and Canada to agree to work together to build a truly international space station. This project was first announced in 1993. This image shows the early stages of the project. The U.S.-built Unity node (right) and the Russian-built Zarya were joined during a December 1998 mission.

Top left: **1972**
On September 5, Arab guerillas of the Black September movement break into the Munich Olympic village, kill two members of the Israeli Olympic team and take nine others hostage. Eventually all the hostages are murdered after a pitched battle at Munich Airport. The event marks an escalation in the war between Arab guerillas and Mossad, the Israeli intelligence service.

Bottom left: **1974**
ABBA win the Eurovision Song Contest for Sweden with their song "Waterloo." Their career is launched into the spotlight following their win and the group go on to become one of the most successful bands of their time.

Top right: **1973**
October 17 – members of OAPEC (the Organization of Arab Petroleum Exporting Countries), cut off exports of petroleum to the U.S. and its allies in Western Europe that have supported Israel in its conflict with Syria and Egypt. The Oil Crisis results in a dramatic increase in oil prices throughout the Western world. On several Sundays in 1973, German highways are empty, as the use of the car is banned.

Bottom right: **1974**
U.S. President Richard Nixon resigns on August 4 in the face of likely impeachment by the United States House of Representatives, as a result of the infamous Watergate scandal.

1970–1979

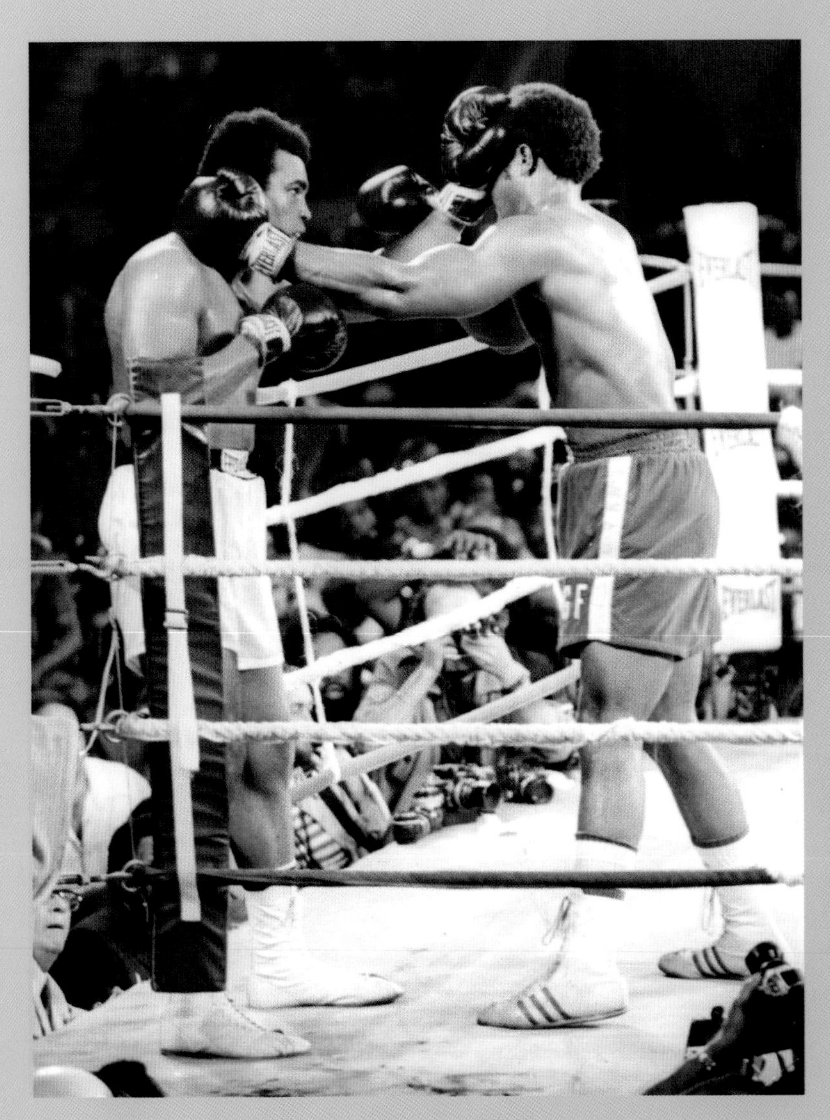

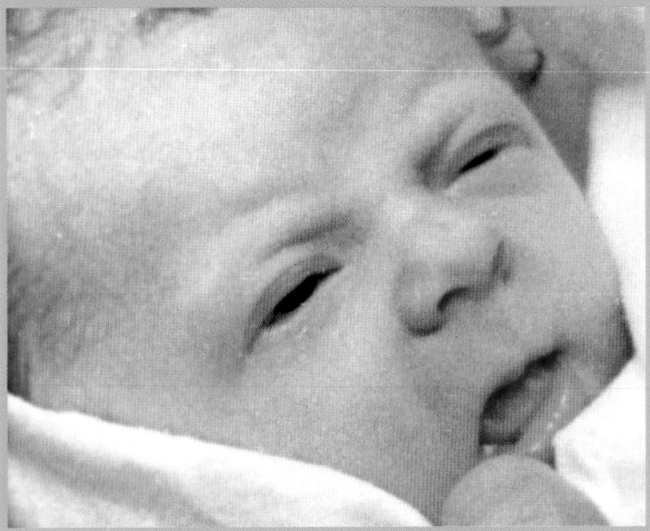

Above: **1974**
Historic boxing event, the "Rumble in The Jungle" takes place on October 30 in Kinshasa, Zaire. The fight pits world heavyweight champion George Foreman against former world champion and challenger Muhammad Ali, who is looking to become the second fighter ever, after Floyd Patterson, to recover the world heavyweight crown.

Top right: **1976**
January 21 – The first commercial flights of the supersonic passenger airliner Concorde take place with a British Airways flight from London Heathrow to Bahrain and an Air France flight from Paris to Rio.

Center right: **1978**
On January 25, the world's first test-tube baby, Louise Joy Brown, is born by Caesarian section at Oldham General Hospital in Lancashire, UK.

Bottom right: **1979**
March 28 – the Unit 2 nuclear power plant on Three Mile Island, Pennsylvania, U.S., suffers a partial core meltdown. It is the worst accident in U.S. commercial nuclear power generating history. Although the accident leads to no injuries or deaths, the cleanup, which starts in August 1979, goes on until December 1993, costing around U.S. 975 million dollars.

Opposite: *Apollo 12*: The second moon landing, November 1969. Charles Conrad photographs Alan Bean, who is collecting samples of lunar material. Note the scene reflected in his helmet and the intense blackness of the lunar sky.

04 ROCKETS TO THE MOON

When the V2 made its first successful flight from Peenemünde, Wernher von Braun made one very significant comment: "It landed on the wrong planet." He had no doubt that the moon was within reach, but several vitally important questions had to be answered before any serious attempt could be made. One of these concerned the nature of the moon itself.

The Earth-facing side had been reasonably well mapped, with its mountains, its valleys, its craters, and its waterless seas—but was it firm? According to some astronomers the answer was "no," so that any spacecraft unwise enough to land there would promptly sink out of sight into a deep layer of soft, treacherous dust. Few practical observers had any faith in this idea, but it did have considerable support, particularly in America. The only way to find out was to make a controlled landing with an unpiloted rocket, but during the 1950s the NASA rockets were anything but reliable. Several attempts were made to target the moon, but the rockets failed to achieve escape velocity, and the first success came from the Soviet Union. On January 2, 1959, Luna 1 (alternatively known as Lunik 1), bypassed the moon at 3,700 miles (6,000 km) and sent back signals for over 60 hours after launch. It was very small, and carried no camera, but it sent back some useful information; for instance, it told us that the moon has no measurable magnetic field—your pocket compass will not work from there!

Next, on September 12, came Luna 2, which achieved a notable first by crashlanding on the lunar surface, probably somewhere near the crater Archimedes in the Mare Imbrium, one of the most conspicuous of the lunar seas (it is bound in part by the Apennines and the Alps). It was followed by Luna 3, which began its journey on October 4, 1959, exactly two years after the real start of the space age. The spacecraft was sent on a trip round the moon, swooping down to less than 4,000 miles (6,500 km) above the surface, and sending back pictures of the regions that are always turned away from the earth. Curious theories had been put forward from time to time; it had even been suggested that the moon was gravitationally lopsided so that all the air and water had been drawn round to the far hemisphere, which might even be inhabited! The Luna 3 images, blurred though they were by the

MAIN MOM
TALK

standards of today, were at least good enough to dispose of these intriguing ideas. The far side of the moon is just as barren, just as crater-scarred, and just as lifeless as the regions we have always known. Yet there are definite differences, and in particular there are no major "seas" apart from the Mare Orientale, a small part of which can been seen from Earth under the most favorable libration. These differences must surely be due to the fact that the axial rotation has been synchronous since a very early stage in the history of the Earth–Moon system.

Other Soviet missions followed; some failed, others were partially successful. On February 3, 1966, came a real triumph: Luna 9 came down in the Oceanus Procellarum, or Ocean of Storms. During approach it had been slowed down by retro rockets, and landed gently; it showed no inclination to sink into dust, and sent back pictures of a scene that looked remarkably like Icelandic lava. Obviously there might be softer areas, but it seemed that the surface in general was strong enough to bear the weight of a spacecraft; remember, too, that on the moon the gravity is only one sixth of that on the surface of the earth. Had the deep-dust theory been correct, manned landings would have been out of the question. NASA planners heaved sighs of relief. In America, things were improving, and after President Kennedy's announcement that the first manned missions were to be sent before 1970, lunar research was given absolute priority. Above all, the Russians were to be beaten in the race to the moon. Finance was not a problem at that stage; in NASA's most expensive year the money spent by the United States on space research was less than that spent on military intelligence alone. Only later did the accountants begin to make their presence felt.

The nine *Ranger* probes (1961 to 1965) were designed to crashland on the moon. They had no hope of survival, but would send back pictures during the last few minutes of their careers. The last three were successful; on March 24, 1965, *Ranger 9* returned nearly 6,000 high-quality images of the large crater Alphonsus, one of a chain of three major walled plains near the center of the moon's disc as seen from Earth (the other two members of the group are Ptolemateus and Arzachel). The seven Surveyor probes (1966 to 1968) were soft-landers; all but two of them worked well, and the deep-dust theory was finally laid to rest. Even more important were the five successful Orbiters (1966 to 1968), which were put into paths round the moon and sent back detailed images of the entire surface—both the Earth-facing and the far hemispheres. All Earth-based maps were instantly obsolete. When the Orbiters had completed their tasks they were deliberately crashed on to the lunar surface so that they could not endanger future missions.

The stage was set for *Apollo*. As most people know, the climax came in July 1969, when Neil Armstrong and Buzz Aldrin reached the moon, but before discussing *Apollo* it may be best to sum up what the Soviets had been doing. The Cold War was at its height, so that there were no contacts between the two teams. The Russian "chief designer," Sergei Korolev, died prematurely during what was expected to be a routine hospital operation; he was probably just as important to the Soviet program as von Braun was to NASA's. There was a disastrous explosion, in which several of the leading experts were killed, and to all intents and purposes the race to the moon was over. After *Apollo 11* the Russians gave up, and concentrated upon unmanned probes, of which the most notable were the two rovers, *Lunokhod 1* (1970) and *Lunokhod 2* (1973). These were taken to the moon by Luna vehicles and crawled around for months, sending back data as well as pictures. *Lunokhod 1*, in the Mare Imbrium, operated for almost a year, travelling over 6 miles (9.7 km) and returning 25,000 images. No. 2, near the incomplete crater Le Monnier, was almost as productive. Both are still on the moon, no doubt waiting to be collected by future astronauts and transferred to a lunar museum. They will come to no harm; there is no corrosion on the airless moon, and there are no storms or cyclones. Ground tremors do occur but are much too mild to be in any way dangerous. Since 1976 the Russians have sent no vehicles to the moon; so far as space is concerned, they have been too busy with other projects.

Apollo, too, met with disaster at an early stage in the program—not in space, but on the ground. During what should have been an ordinary routine test the capsule caught fire. The three astronauts inside had no chance of escape. *Apollo 7* was the first to fly, in October 1968; then, in the following December, Frank Borman, Jim Lovell, and Bill Anders went round the moon in *Apollo 8*, and became the first men to see the "other side." *Apollo 9* (March 1969) was another Earth-orbiter, but *Apollo 10* (May 1969) did everything except touch down on the moon. The astronauts—Tom Stafford, John Young, and Eugene Cernan

Opposite: Computer artwork impression of Luna 1 passing the moon, October 1959. Whether or not it was intended to land has never been cleared up. In the event it flew past at 3,700 miles (5,955 km) sending back valuable data. Presumably it is still in orbit around the sun.

made a careful survey of the site chosen for the first actual landing, near the edge of the Mare Tranquillitatis—the Sea of Tranquillity.

MEN ON THE MOON

I think that almost every educated person will know that Neil Armstrong was the first man to reach the moon, and most will know the name of his companion Buzz Aldrin—but how many remember Michael Collins, the third member of the crew of *Apollo 11*? Yet his role was absolutely vital, and without his help the landing would have been impossible. (As he later commented, he went 99 percent of the way.) The three travelled in the main spacecraft and went into orbit round the moon; Armstrong and Aldrin then entered the Lunar Module, "Eagle," and made the final descent. Bear in mind that there was no chance of rescue in the event of a faulty landing, and it was a tremendous relief, to put it mildly, when Neil Armstrong's voice came through: "The Eagle has landed." The gap between the two worlds had been bridged. A few hours later, the two Earthmen were standing on the lunar rocks.

This is how Neil Armstrong described the scene:

> You generally have the impression of being on a desert-like surface, with rather light-colored hues. Yet when you look at the material from close range, as in your hand, you find that it's really a charcoal gray. We had difficulties in perception of distance. For example, from the cockpit of the lunar module we judged our television camera to be only 50 or 60 feet away, yet we knew that we had pulled it out to the full extent of a 100-foot cable. Similarly we had difficulty in guessing how far away the hills on the horizon might be from us. This peculiar phenomenon is the closeness of the horizon, due to the greater curvature of the moon's surface—four times greater than on Earth; also its an irregular surface, with crater rims overlying other crater rims.

This is a changeless world; nothing breathes, nothing stirs; there can be no natural sound. The moon had never known life before the astronauts arrived. The stark landscape has a strange beauty; no one has bettered Buzz Aldrin's description of it as "magnificent desolation," with the sun shining down from a jet-black sky. Stars cannot be seen, because of the glare from the lunar surface, but the earth is there 250,000 miles (402,336 km) away. The first moonwalkers did not go far from the grounded Eagle, their only link with home, but they were very busy; they collected samples of the rock, and then set up the ALSEP, the Apollo Lunar Surface Experimental Package, which included a variety of instruments—for example, a solar wind detector, and a seismometer to record any "moonquakes." There were no communication problems, and neither were there any problems about moving around under one sixth of the earth's gravity; everything seemed to happen in slow motion.

All in all, the whole mission was remarkably trouble-free. The next really tense moment came when Armstrong and Aldrin were back inside the Lunar Module, ready to blast off to rejoin Collins in the orbiting Command Module. The lunar lander had only one engine—and this had to work, perfectly, first time. There could be no second chance. Mercifully, there were no mishaps. Using the bottom part of the Module

Above: A NASA *Surveyor* spacecraft identical to the one which made the first U.S. soft landing on the moon on June 1, 1966, is tested back on Earth by technicians.

Opposite: A mosaic, 30 square feet (2.8 square meters), spread out on the floor at NASA headquarters. It was assembled by the U.S. Army Map Service in 1968, and is made from 127 photographs taken by Orbiter 4 in 1967.

as a launch pad, the top part fired back into orbit, and made rendezvous with the patient Collins. On the morning of July 22, *Apollo 11* was put into an Earth-directed orbit, and splashed down in the Pacific at 4:49pm on July 24. After a total of 195 hours in space, the splash down—less than 12 miles (19 km) from the intended position—was a mere thirty seconds late.

Much had been learned. As expected, the rocks were of volcanic type, mainly basaltic; there was no trace of any hydrated material, so it seemed that the moon must always have been bone dry; the lack of atmosphere was confirmed, and so too was the absence of any overall magnetic field; a reflecting mirror was set in place, to be contacted by laser beams from Earth and making it possible to measure the moon's distance to an accuracy of less than 1 in. (2.5 cm). The ALSEP instruments went on working for some time after the astronauts had left.

There was one other point. Though it seemed virtually certain that there were no harmful substances on the moon, one could not be absolutely sure (remember Professor Quatermass!) and so the returning astronauts were quarantined on return until they had been thoroughly checked. Quarantining was abandoned after the first two missions because it was clear that there really was no risk, but more stringent precautions will have to be taken when astronauts come back from Mars, which probably supported life in past ages and may still do so today.

In December 1969, *Apollo 12* carried Charles Conrad, Alan Bean, and Richard Gordon to the lunar Oceanus Procellarum. They came close to the old unmanned probe *Surveyor 3*, which had been on the moon ever since 1967. Alan Bean walked across to it, chipped small pieces off, and brought them for analysis; predictably they were undamaged.

Above left: The *Apollo 11* crew. Left to right: Neil Armstrong (commander), Michael Collins (Command Module pilot), and Edwin (Buzz) Aldrin (Lunar Module pilot). Armstrong and Aldrin walked on the moon; Collins said at least he went "99 percent of the way."

Above right: Buzz Aldrin descending the ladder to join Armstrong on the lunar surface. The photograph was taken by Armstrong, who was responsible for almost all of the photography—which is why there are so few pictures of him and so many of Aldrin.

In all essential ways the results from *Apollo 12* confirmed those of *Apollo 11*, and there were no real surprises.

When *Apollo 13* took off from Cape Canaveral, on April 11, 1970, it did so without quite the fanfare of publicity which had marked the earlier missions; there seemed to be a feeling that lunar travel would soon become a matter of routine. But near disaster lay ahead. The original crew consisted of James Lovell, Fred Haise, and Ken Mattingly; at the last moment it was found that Mattingly had been exposed to German measles, and he was replaced by Jack Swigert. (In fact, he did not develop German measles and later flew in *Apollo 16*.) During launch, one of the motors cut out early, but the others functioned according to plan, and *Apollo 13* was on schedule as it went on its way. Conventional reports to Mission Control at Houston were followed by a television program from the capsule, watched by millions of people. Then, with *Apollo* 178,000 miles (286,000 km) from Earth, came the message from Lovell, which must surely be the understatement of the century: "Houston, we've had a problem." There had been an explosion, putting the service module permanently out of action. Henceforth, the only concern was to bring the astronauts home alive.

All thoughts of a lunar landing were abandoned at once, since the only engine which could be used were those of the Lunar Module ("Aquarius"). It was impossible to turn the module around and bring it straight back to Earth; the only hope was to go around the moon and then use Aquarius' ascent engine to accelerate the whole spacecraft and bring it back to a suitable splash down point. Fortunately the plan worked, and the story of *Apollo 13* had a happy ending—but it had been touch and go.

Apollo 13 had been scheduled to land near three low-walled craters in the Mare Nubium, or Sea of Clouds: Fra Mauro, Bonpland, and Parry. *Apollo 14* (February 1971) was programmed to go to the same region, and this time there were no major hitches. One of the two moonwalkers was Alan Shepard, who had become America's first man in space a decade earlier; his companion was Edgar Mitchell. With them they took a "cart," which they dragged with them on an extended moonwalk: they set up an ALSEP, collected samples and took photographs. Just before reentering the lunar module, Shepard took

Top right: The Lunar Module of *Apollo 11* on its way back to the Command Module after the two moonwalkers had blasted away. The picture was taken by Collins from the Command Module itself.

Bottom right: *Apollo 11* view of the full moon, taken during the spacecraft's trans-Earth journey. When this picture was taken, the spacecraft was already 11,516 miles (18,533 km) away.

time off to tee up a golf ball and drive it; apparently it travelled several hundred meters—ever since then, sports gear manufacturers have been trying to find out what type of ball was used! (Years later, a Russian cosmonaut was scheduled to drive a ball from a platform attached to the *International Space Station*, but this golf ball, unlike Shepard's, was to be fitted with a radio transmitter, and was expected to orbit the earth for several years before dropping back into the atmosphere and burning away in the same manner as a meteor.)

Apollo 15 (July 1971) was sent to the most spectacular site yet; the region of the Hadley Rill, in the foothills of the Apennine mountains. The Apennines, bordering the Mare Imbrium (Sea of Showers) have peaks rising to over 15,000 feet (4,572 m) above the surrounding terrain, and to an Earth-based observer make up the most conspicuous range on the entire moon; the Rill runs parallel to the edge of the Mare, and for much of its length is about 1 mile (1.6 km) wide and 1,300 feet (396 m) deep: large rocks have rolled down on to its floor.

The moonwalkers were David Scott and Jim Irwin, with Alfred Worden staying in the orbiting section of the spacecraft. The crew took a moon car with them. Known as the LRV or Lunar Roving Vehicle; it was electrically driven, and could race along at the giddy speed of 8 mph (13 km/h). It looked fragile, and had to be folded up during the journey to the moon, but it worked well, and it meant that the astronauts could travel in relative comfort; altogether Scott and Irwin covered a distance of over 25 miles (40 km). The main emphasis from a scientific point of view was on geology. The second moon drive took the astronauts to the base of the mountain known as Hadley Delta, one of the loftier peaks of the Apennines, where they found the lemon-sized Genesis Rock, which is thought to date back 4.5 thousand million years—in fact, almost to the time of the moon's birth. In the official catalog, the Genesis Rock is listed as Apollo Sample 15415. Altogether the astronauts collected 34 lbs. (76 kg) of samples for analysis.

Incidentally, Hadley Delta was quite unlike the jagged, steep-sloped peaks shown in old artists' impressions and science fiction films. Scott even described it as "a featureless mountain," and the slopes were fairly gentle. Future mountaineers should have little difficulty in climbing them, in spite of the need to wear vacuum suits.

Above left: *Apollo 12*: Charles Conrad photographs Alan Bean on the Oceanus Procellarum, close to the *Surveyor 3* spacecraft that had landed there in 1967. Note the glare of the sun; the camera was aimed rather too close to the sun itself!

Above right: The damaged *Apollo 13* Service Module, photographed from the Command Module after the Service Module had been jettisoned. An entire panel has been blown away. Had the explosion occurred on the return journey, the Lunar Module *Aquarius* would have been jettisoned and nothing could have saved the astronauts.

Opposite: Lift-off of *Apollo 17*, December 7, 1972, carrying astronauts Cernan, Schmitt, and Evans: the last of the *Apollo* missions. This was a night launch, which made it all the more spectacular for onlookers.

Apollo 16 (April 1972) came down in the lunar highlands, near the crater Descartes, and the astronauts, John Young and Charles Duke, carried out a full scientific program. But probably the most productive of all the *Apollos* was no.17 (December 1972), which touched down in the region of the Taurus Mountains, in a valley near the incomplete crater Littrow. (The Taurus Mountains consist of separate clumps of hills and do not make up an Apennine-type range.) The commander was Eugene Cernan; the other moonwalker was Harrison Schmitt, a professional geologist who had been given astronaut training especially for the mission. His specialized knowledge proved to be invaluable.

"It's orange—crazy!" There was great excitement at Mission Control in Houston because it was thought that the color might indicate fumarole activity—in fact, ongoing vulcanism. Alas, it turned out to be due to swarms of colored beams, shot out during an explosive volcanic eruption of the fire-fountain type, and they were around 3,640 million years old—much more ancient than Shorty itself, which was an ordinary impact crater.

During their 22 hours outside the Lunar Module, Cernan and Schmitt collected over 240 lbs. (109 kg) of samples; they took many photographs, and made scientific observations of all kinds. When they reentered the module, their work had been well done. Cernan was actually the last to leave—it would be a long time before anyone else set foot on the moon. Asked later what had impressed him most, he gave a very definite answer: "Seeing my home world, so far away." When the module blasted away, the first chapter of man's exploration of the moon had come to an end.

LUNAR ICE?

Plans had been made to send three more missions, but *Apollo 18, 19,* and *20* never progressed beyond the drawing board. There were several reasons for this. The accountants were becoming more and more vocal, and the international situation was still very unsettled, but it was also realized that *Apollo* had achieved everything that had been hoped of it; extra missions would not achieve a great deal, and sooner or later something would have gone badly wrong with tragic consequences. So *Apollo* was consigned to history, and for more than two decades the moon was left unvisited apart from four Russian unpiloted probes, one of which carried the "crawler" *Lunokhod 2.*

Left: On the moon with *Apollo 17*. Cernan and Schmitt, driving in the Taurus-Littrow area, photograph the grounded Lunar Module; the tracks of their Lunar Rover are very obvious, but do not sink deeply into the surface material.

Below: Harrison (Jack) Schmitt, with *Apollo 17*, collecting lunar samples. Schmitt is the only geologist to go to the moon; he was given astronaut training specially for the mission, and his expertise proved to be of the greatest value.

The lull ended on January 2, 1994, with the launch of Clementine, named after the character in the old mining song, who is "lost and gone forever." It was given a complicated program. It would be put into a circumlunar orbit on February 24, and would begin a mapping program with special emphasis on the polar regions: it would then break free, and go on to rendezvous with a small but interesting asteroid, Geographos. In the end, things did not go according to plan because a fault developed and the Geographos rendezvous had to be abandoned, but so far as the moon was concerned, Clementine performed well, and at one point was only 260 miles (420 km) above the lunar surface. The images sent back were excellent, and one of them showed the South Pole—Aitken Basin, the largest and deepest on the moon—over 1,500 miles (2,500 km) across and with a floor 8 miles (13 km) below the adjacent landscape. But there was one very unexpected development. On December 3, 1996, the Pentagon announced that ice had been discovered on the floors of some of the polar craters, which are always in shadow and therefore remain bitterly cold. It was claimed that if all the ice were melted it would fill a lake 2 miles (3 km) square and 3 ft. (1 m) deep!

This sounded exciting; if confirmed it would make the moon a much less hostile world, a revelation which would go some way toward placating the accountants.

There was no suggesting that Clementine had found a sheet of ice; there are definitely no skating rinks on the moon! The ice was said to be mixed in with the other materials, so that extracting it would be far from straightforward. The discovery—if it really were a discovery—had been made with an instrument known as a neutron spectrometer. When cosmic ray particles hit the moon, neutrons are produced, moving at very high speeds. However, if a neutron happens to hit a hydrogen atom it is slowed down, because it loses some of its momentum—just as a snooker ball will do after cannoning off another ball. Collisions with heavier atoms do not have the same effect, so if we find a large number of "slow" neutrons it is safe to assume that hydrogen is present. As Clementine passed over the polar craters it recorded more slow neutrons than would usually be expected. It was then claimed that the hydrogen had combined with oxygen to form water.

It all sounded rather glib, but NASA remained confident. The next probe, *Prospector* (January 1998), seemed to confirm Clementine's results and one NASA scientist, William Feldman, stated categorically that "we have found water," perhaps between 2,600 million and 80,000 million gallons (10,000 million and 300,000 million liters) of it; "a significant resource which will allow a modest amount of colonization for many years. Water can now be mined directly on the moon, instead of having to be shipped from Earth."

There were plenty of skeptics; how could the water have got there? Samples brought home by *Apollo* and the Russian probes had shown no sign of hydrated materials; it could hardly have been deposited by a comet, because it would have been blasted away rather than left to condense and the hydrogen could well have come from the constant stream of solar wind. On July 31, 1999, *Prospector* was deliberately crashed in the polar zone in the hope that water would be detected in the debris thrown up. The results were negative, and although the search will go on, the existence of lunar ice now seems very unlikely—much to NASA's disappointment.

The latest lunar probe (to date) has been the European *SMART-1*, launched on September 27, 2003, by an Ariane rocket from the Guiana Space Centre. It was put into a "parking orbit" round the earth, and then used ion propulsion to send it moonward. The journey took over a year, but the method worked perfectly, and *SMART-1* undertook an extended program of mapping. Ion drives will become more and more important as time goes on.

We have found out more than would have seemed possible at the start of the space age, and the idea of a fully-fledged lunar base is no longer far-fetched. There is every reason to hope that a base will be established during the first half of the present century; it will be international, and immensely valuable to all mankind. Life will have come at last to the faithful, waiting moon.

Opposite: At the end of a long day on the moon, *Apollo 17* astronaut Gene Cernan rests inside the Lunar Module *Challenger*. Note the moondust covering his suit. Cernan described the dust as smelling like "spent gunpowder," and his fellow astronaut Jack Schmitt found it gave him hayfever.

Below: This mosaic is composed of 1,500 Clementine images of the south polar region of the moon. Clementine has revealed what appears to be a major depression near the lunar south pole (center), probably an ancient basin formed by the impact of an asteroid or comet. A significant portion of the dark area near the pole may be in permanent shadow, and sufficiently cold to trap water of cometary origin in the form of ice.

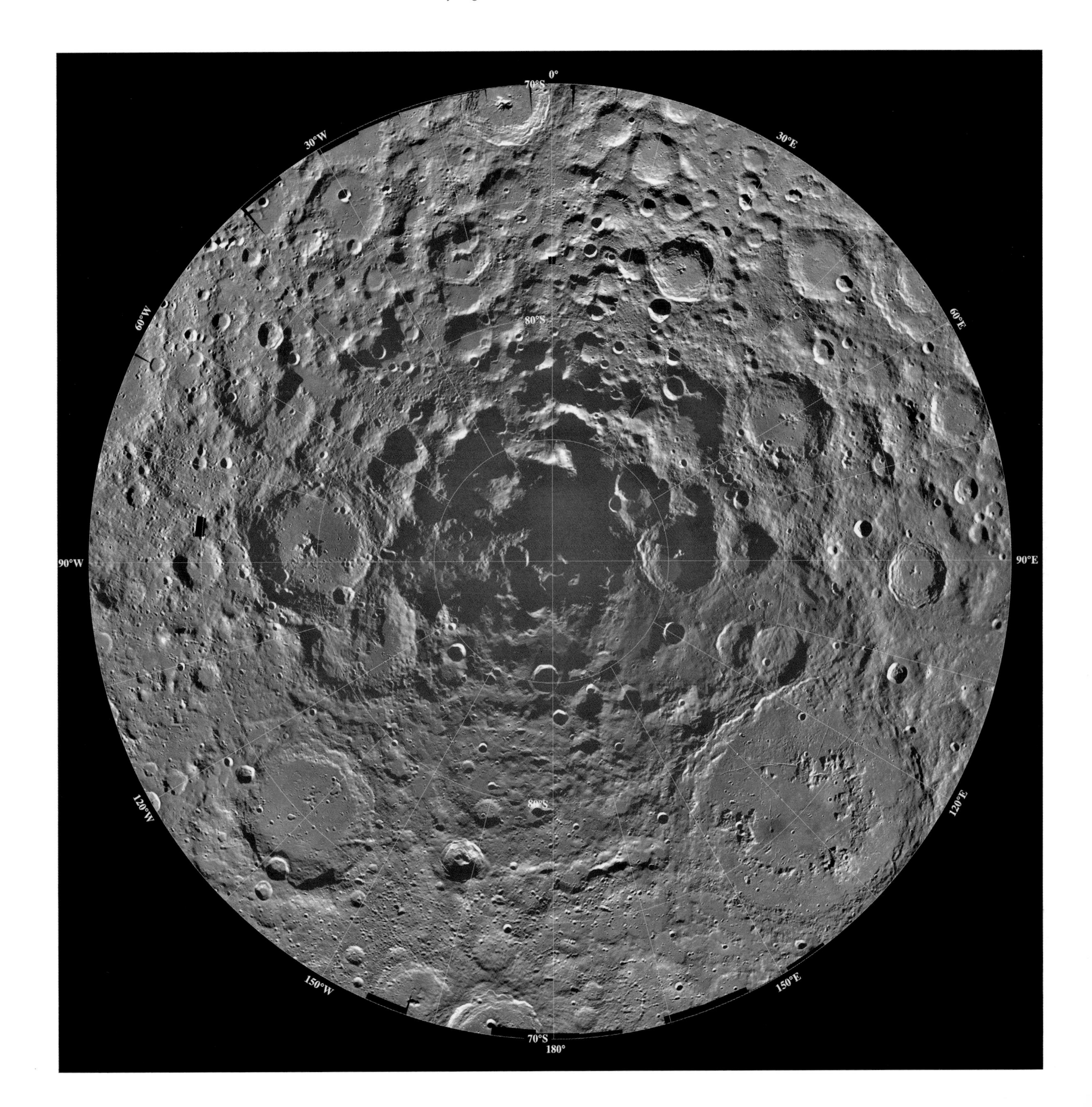

Opposite: The planet Venus is dwarfed by the sun as it crosses its face in 2004, an event known as a transit. The dark disk of Venus seems to be imitating a giant sunspot that looks perhaps a little too round.

05 NEIGHBORS

In the solar system there are only three planets which may be regarded as being within what may be called "reasonable" range. These are Mars, Venus, and Mercury. Therefore, it was fairly obvious that these would be our first space targets after the moon. Mars seemed to be particularly promising, and it was generally believed that Venus would not be too unlike Earth, while Mercury would not be too unlike the moon. We were right about Mercury, but about Venus we could not have been more wrong!

TO VENUS

The problem with Venus was that we could not see the actual surface, which is hidden by the thick, cloud-laden atmosphere. Venus is closer to the sun than we are— 67 million miles (108 million km) out, as against our 93 million miles (150 million km)—and moreover we could use spectroscopic analysis to show that the planet's upper atmosphere, at least, was rich in carbon dioxide, which is a "greenhouse" gas, and tends to shut in the sun's heat. Venus was clearly a hot place—but how hot? There were plenty of theories. Venus might be a tropical world, similar to Earth at the time when the coal measures were being laid down, with swamps, luxuriant vegetation, and huge dragonflies; it might be a hot desert; it might be mainly ocean-covered; Fred Hoyle even suggested that there could be seas of oil. We knew that the orbital period—Venus' "year"—was almost 225 Earth days, but we did not know the length of the axial rotation period. Venus was often called "the planet of mystery." The only way to learn more was to send a spacecraft.

At its nearest, Venus can come within 24 million miles (39 million km) of earth: about 100 times the distance from Earth to the moon. Unfortunately, we cannot simply wait until Venus is closest, and then just fire a rocket across the gap. This would mean using more propellant than any spacecraft could carry, and we have to use a different method. The probe is launched and put into an Earth orbit. At a suitable moment it uses its own rocket motor

to put itself into a new path which will take it to Venus without using any more fuel; it is "coasting," so to speak, by making use of the sun's gravitational pull. The journey is bound to take months but so long as we have to depend on chemical propellants, there is no short cut.

The Russians took the lead. On February 12, 1961, they launched *Venera 1*, and at first all seemed well, but a week later contact was lost and was never reestablished, so we will never know quite what happened. In the United States, NASA was also eyeing Venus, and began their Mariner program. *Mariner 1* (July 22, 1962) was not a success; it left its launch pad and dived into the sea—someone had forgotten to feed a minus sign into a computer, with predictable results. However, *Mariner 2*, dispatched a month later, more than made up for the sad fate of its predecessor. It carried no camera, and there was to be no landing: *Mariner 2* bypassed Venus in December at a range of 21,594 miles (35,000 km), and remained in touch until January 4, 1963. What it told us made us realize that although Venus and the earth were so alike in size and mass, they were very dissimilar twins.

First, the surface temperature turned out to be about 800°F (430°C), which at once ruled out the idea of watery oceans as well as Fred Hoyle's oil wells. Secondly, the rotation period was very long—we now know it to be 243 Earth days, longer than the orbital period, producing a very peculiar calendar; moreover, Venus spins from east to west, so that it may be regarded as an "upside-down" planet. From the surface the sun would rise in the west and set in the east—though from the surface the sun would never be seen through the cloud; there is no such thing as a sunny day on Venus. The atmosphere consists mainly of carbon dioxide all the way through, not only in its upper layers. And there is no detectable magnetic field, so presumably there is no iron-rich core like that of the earth. All in all, Venus was clearly unwelcoming, which was probably why, for the next few years, NASA transferred its attention to the less hostile Mars.

However, the Russians remained enthusiastic about Venus. Their stated aim was to bring a spacecraft down gently, so that it could transmit direct from the surface, but there were many unknown factors; if Venus was really covered with ocean, a probe would have to be prepared to float, and it would certainly have to be chilled before beginning its descent through the dense, scorching atmosphere. *Venera 2* (1966) missed Venus altogether, but *Venera 3* probably landed on March 1, 1966. Disappointingly, contact was lost before touch down. The density of the lower atmosphere was greater than had been expected, and the luckless *Venera* was literally squashed. *Veneras 4*, *5*, and *6* suffered the same fate. The first success came with *Venera 7*, which was launched on August 17, 1970, and landed on the following December 15, at the east end of the area now known as Navka Planitia; it managed to transmit for 23 minutes before succumbing to the intensely hostile environment. *Venera 8* (July 1972) did even better and transmitted for 50 minutes before falling silent. The high surface pressure and temperature were fully confirmed, and the illumination was found to be "about the same as that in Moscow at noon on a cloudy winter day." The wind speeds seemed to be very low, and this was a surprise, because the winds in the higher atmosphere were known to be strong. We now know that the upper clouds are quick-moving, with a rotational period of only 4 days—a remarkable contrast with the slow spin of the planet itself. Just why the atmosphere has this curious structure is unclear, and neither do we know why Venus turns from east to west. It has been suggested that in its early stages a collision with a large asteroid caused Venus to "tip over," and though this sounds unlikely it is difficult to think of any better explanation.

There were two NASA missions during this period; *Mariner 5* (1967), a fly-by, and *Mariner 10* (1974), which passed within range of Venus on its way to rendezvous with Mercury, but the next striking development came in October 1975, when *Veneras 9* and *10* made controlled landings within a week of each other (October 21 and 25 respectively). Each sent back one picture, and each remained in touch for around one hour. For the first time we could really see what the landscapes of Venus were really like.

Opposite: The ill-fated *Mariner 1* was intended to perform a Venus fly-by but was destroyed 293 seconds after launch when it veered off course, possibly heading for the North Atlantic shipping lanes or an inhabited area. A tiny mistake in the flight program was found to be the cause of the problem.

Above: Venus from *Mariner 10* on February 6, 1974 (ultraviolet light, showing cloud features). *Mariner* was on its way to rendezvous with Mercury, and had flown past Venus at a range of 3,600 miles (5,794 km) one day before this image was taken.

The pictures were amazingly good. *Venera 9*, near the small highland area known as Beta Regio, showed a rock-strewn scene; the rocks were sharp-edged, and most of them were 2–4 ft. (60–120 cm) across. It was said that we were looking at "a stony desert"; there was no obvious sign of erosion. Measurements indicated a windspeed of about 8 miles per hour (13 km/h); this may sound gentle but in Venus' thick atmosphere it would be as devastating as the effect of a sea-wave on a terrestrial cliff during a 46 miles per hour (74 km/h) gale, so that the lack of erosion was another surprise. The scene from *Venera 10* was much the same, but it looked as though the larger rocks were outcrops from below the surface rather than a simple scattering. Also, the rocks were smoother, and the terrain appeared to be older than that of the *Venera 9* landing site a few tens of kilometers away.

NASA turned its attention back to Venus in 1978, with *Pioneers 12* and *13*. The first of these was an orbiter, but the second (often referred to as *Pioneer Venus 2*) was much more complicated. It too was aimed at the area of Beta Regio and was a multi-probe arrangement, consisting of a "bus" transporting three landers. The landers were separated from the bus in November, and came down independently; their main function was to send back data during their final descent, and they were not designed to survive impact on the surface, although one of them actually did so, and transmitted for 67 minutes after arrival. The bus had no braking mechanism, and simply crashed into the area of Themis Regio. (Of course, these various regions had been identified by radar mapping, some of which had been done by Earth-based equipment; the atmosphere of Venus is no barrier to radar pulses.) The results showed that the clouds, far from being watery, are in sulphuric acid. As a world, the Planet of Love was looking more and more unattractive!

Several more missions followed during the 1980s; three *Veneras*, and two Soviet probes, *Vega 1* and *Vega 2*, which paid courtesy visits to Venus on their way to welcome Halley's comet, which was returning to the inner part of the solar system for the first time in 76 years. As they passed by, in June 1985, each *Vega* dropped a balloon into Venus' atmosphere and each deposited a small lander. The balloons drifted around and sent back information about windspeeds and temperatures; when they moved into the hot atmosphere on the day side of Venus they expanded and burst. The landers transmitted for almost half an hour after impact.

Next came Magellan, the orbiting radar mapper which was launched on May 5, 1990, from the space shuttle *Atlantis*. It reached its target on August 10, and went on moving around Venus until October 11, 1994, when it fell back into the planet's atmosphere and burned up. It was put originally into an eccentric orbit, bringing it to within 185 miles (298 km) of the surface at closest approach; during its career, the orbit was modified several times. Magellan was an unqualified success. It could resolve features down to less than 400 ft. (120 m) and altogether it surveyed about 98 percent of the total surface.

The mapping procedure followed a set pattern. During the part of its orbit closest to the planet, Magellan's radar mapper imaged a swath

Above: *Venera 7* landed on Venus in 1970, sending back the first radio messages from the surface. Later *Venera* probes took photographs of the planet's surface. The extreme conditions often meant the probes did not survive long.

Opposite: Venus from *Venera 13*, March 1, 1982. The scorching-hot surface is rock-strewn: the light was said to be as bright as that in Moscow at noon on a cloudy winter day. *Venera 13* transmitted for an hour after landing.

Overleaf: Perspective view of the 5-mile (8-km) high shield volcano Maat Mons. The image is based on radar mapping from the Magellan orbiter; color based on *Venera 13* and *14* images. Obvious lava flows can be seen extending for hundreds of miles.

ИППИ АН СССР
НЕРА-13 ОБРАБОТКА И

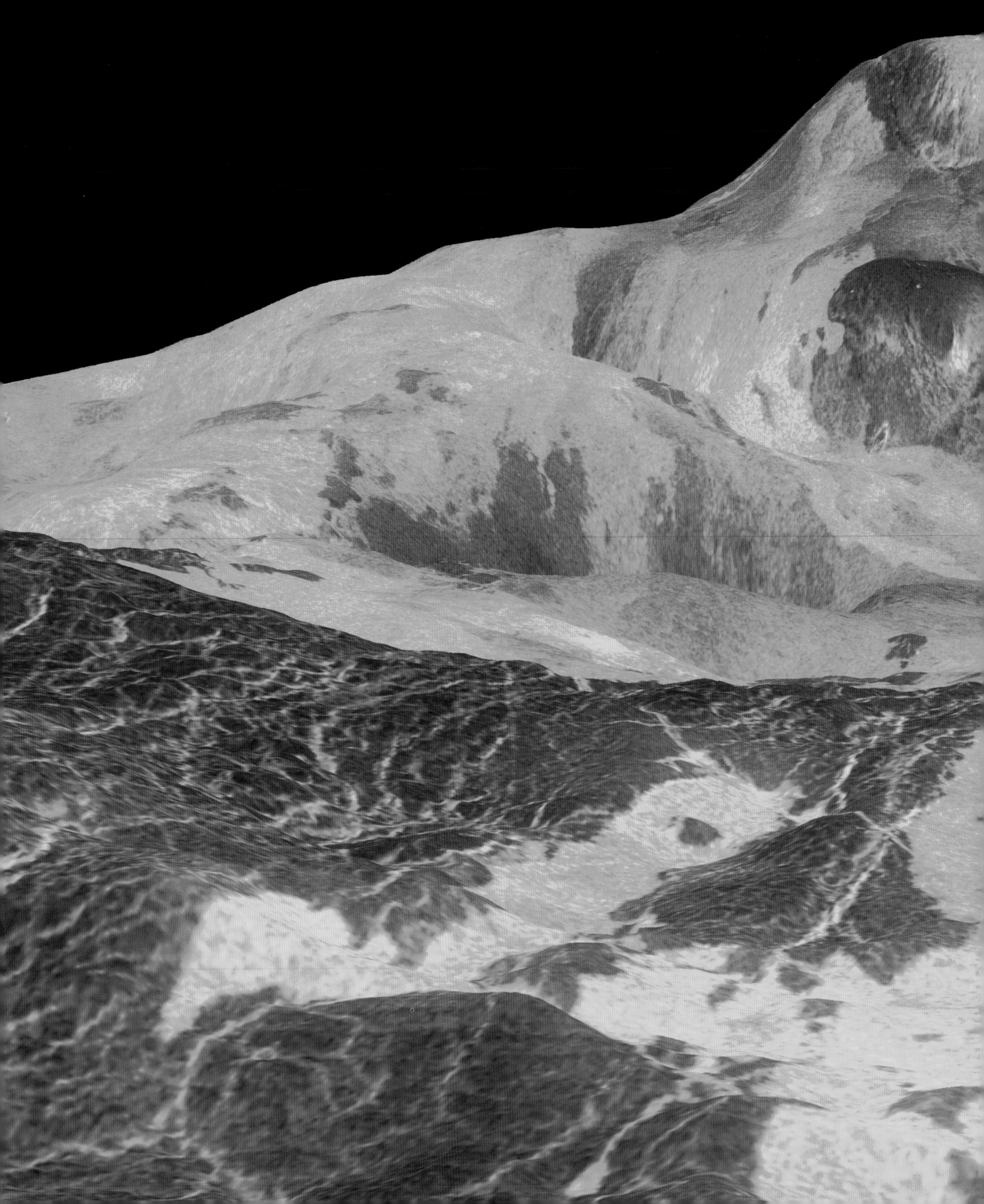

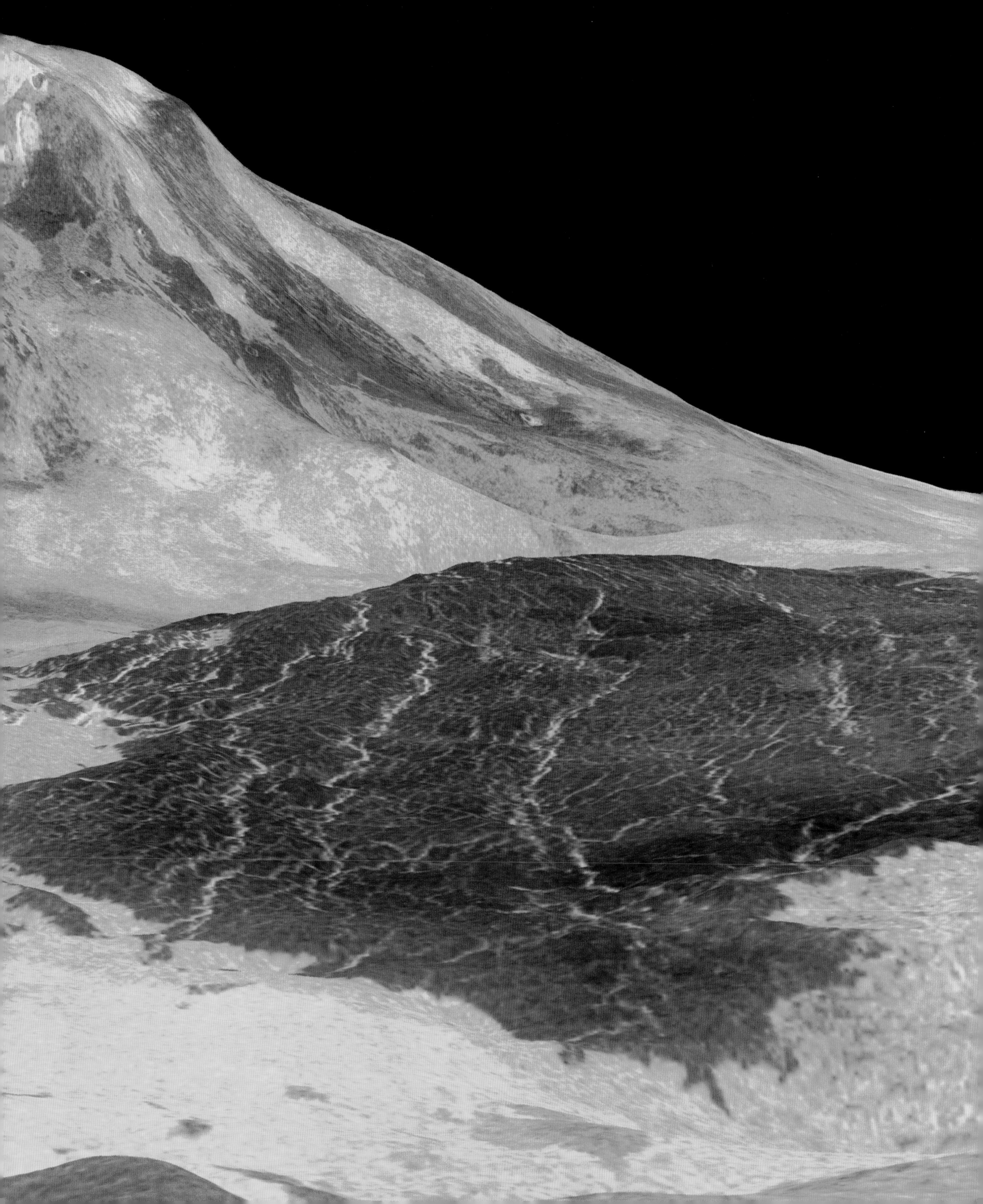

Above: Visible/ultraviolet image of Venus taken from the *Venus Express* spacecraft, April 19, 2006, from a range of about 120,000 miles (193,121 km). On the day side (blue part) stripelike structures can be seen: they were first seen by *Mariner 10* in 1974 and may be due to the presence of dust and aerosols in Venus' atmosphere.

of the surface about 15 miles (24 km) wide. The orbital period was 3 hours 15 minutes. At the end of each orbit, Magellan sent back to Earth a long, ribbon-like map of the strip that had been surveyed. Venus itself has a rotational period of 243 Earth days; as the planet rotated beneath the spacecraft, strip after strip of the surface was imaged, until at the end of the 243 day cycle virtually the whole of the planet had been covered.

A HELLISH WORLD

Venus is a world of volcanic plains, highlands and lowlands. The highlands account for about 8 percent of the surface, the lowlands (planitiae) account for 27 percent and the rolling plains for the remaining 65 percent. Vulcanism dominates the scene, and there are lava flows everywhere; there is every reason to suppose that volcanic eruptions are going on at the present time—there is nothing calm or peaceful about Venus.

Obviously it is convenient to name the various features, and this is the responsibility of the Nomenclature Commission of the International Astronomical Union. In the cause of equality, it has been decreed that all names on Venus must be female—males are represented by James Clerk Maxwell: the highest mountains on Venus were named in his honor before the ban was imposed! Many of the names are familiar—Marie Curie, Florence Nightingale, Kirsten Flagstad—but how many people will know that, for instance, Vellamo is a Finnish mermaid or that Xochiquetzal is the Aztec goddess of flowers?

Thousands of volcanoes have been identified on Venus, ranging from tiny vents up to huge shield volcanoes thousands of metres high, similar in type to the earth's Mauna Kea and Mauna Loa on Hawaii, but much loftier and more massive. Predictably, many of these volcanoes are crowned by large summit calderae. Of special interest are the features known as pancake domes, which are always circular and not particularly high; they may be due to very thick lava-flows breaking through the surface from below.

There are two main upland areas, Ishtar Terra in the northern hemisphere and Aphrodite Terra, which is mainly in the south but is crossed by the equator. Ishtar is about the size of Australia, and consists of eastern and western components, which are separated by the Maxwell Mountains, which rise to 5 miles (8 km) above the adjacent terrain and have steep slopes of up to 35 degrees in places. Maxwell forms the eastern edge of a high plateau, Lakshmi Planum, which is bounded to the west and north respectively by the Freyja Akna and Danu Mountains. Lakshmi is relatively smooth, covered with lava.

Aphrodite Terra is larger—about 3,000 x 2,000 miles (4,828 x 3,219 km)—and is made up of eastern and western upland areas separated by a lower region. Western Aphrodite consists of two highlands, Thetis and Ovda, characterized by what used to be called "parquet" terrain, though this term was abandoned as being unscientific and was replaced by "tesserae"; it is extremely rough and, so far as we know, is unique to Venus.

The third important highland is Beta Regio, south of the equator adjoining Phoebe Regio and Themis Regio. It includes two huge structures: Theia Mons and Rhea Mons. Both were once thought to be volcanoes; this is true of Theia, which is indeed the most massive of all Venus' volcanoes, but not of Rhea, which is however scoured with volcanic deposits. Theia is over 15,000 ft. (4,500 m) high, with a base diameter of 220 miles (350 km) and an oval central caldera measuring 46 x 30 miles (74 x 48 km). Lava-flows from Theia extend over an area more than 500 miles (800 km) wide; the volcano lies at the junction of three rifts, one of which, Davana Chasma, is 125 miles (200 km) wide in places, making our Grand Canyon seem puny. South of Rhea between Beta Regio and Phoebe Regio, the chasm is 4 miles (6 km) deep; its full length is over 1,000 miles (1,600 km). If astronauts ever go to Venus, there is no doubt that Devana will be a leading tourist attraction!

Other tectonic features include coronae, arachnoids, and novae. Coronae are found only on Venus; hundreds have been identified, up to 150 miles (250 km) across. A typical corona is a circular structure, with a circumference defined by a radar ring-like zone made up of troughs and ridges. It is thought that a corona may be due to a mass of hot magma below the surface rising upward and breaking the crust causing it to melt and collapse. Arachnoids, which may be as much as 140 miles (230 km) across, are so named because of their spidery and cobweb-like appearances in radar images. A typical arachnoid has a central volcanic feature surrounded by a network of fractures; in some ways they are not unlike coronae, although they are smaller. Novae have radial structure, and may represent early stages in the formation of coronae.

Magellan was a triumph; its makers were sad when it plunged to its fiery death. Now we have Europe's Venus Express, launched from Baikonur Cosmodrome, in Kazakhstan, on November 9, 2005, atop a Soyuz-Fregat rocket. It was put into a parking orbit, and then sent off on its main journey; it entered its final operational orbit round Venus on May 7, 2006, and began its long program of photography and general research and analysis. Its nominal lifetime was 500 Earth days—approximately two Venus years—and it carried a whole battery of instruments. For example, there was a magnetometer designed to measure the planet's very feeble magnetic field, and a Venus Radio Science Experiment (VeRa), which will transmit radio waves from the spacecraft and pass them through the atmosphere or reflect them off the surface. These radio waves are received by a ground station on Earth to be used for analyses of the ionosphere, atmosphere, and surface details of Venus.

One feature of Venus is that there are not many very small impact craters, simply because small meteoroids cannot survive the drop through the dense atmosphere. However, major impact craters are common; the largest known, Mead (latitude 12.5 degrees N, longitude 57.2 degrees E) is 174 miles (280 km) in diameter, with an inner and an outer ring and a small surrounding ejecta blanket, while the 39 mile (63 km) Alcott (latitude 59.5 S, longitude 62.7 E) has been flooded by lava so that its floor is featureless. There are long lava channels, such as Hecate Chasma, which runs for almost 2,000 miles (3,000 km). The scenery on Venus is nothing if not spectacular!

Venus has no satellite. Careful searches have been made, but without success. A satellite as much as 1 mile (1.6 km) across would certainly have been detected by now. Of all the planets, only Venus and Mercury are solitary travellers in space.

MERCURY

Mercury is never very conspicuous as seen from Earth; its mean distance from the sun is only 36 million miles (58 million kilometers), so that in our skies it always keeps to within 30 degrees of the sun, and is visible with the naked eye only when either very low in the west after sunset or very low in the east before dawn. Earth-based telescopes show little more than the characteristic phase, and for detailed information we rely upon spacecraft. At the time of writing, only one probe, *Mariner 10*, has bypassed Mercury. Another, *Messenger*, is on the way, but will not arrive there until 2011. Like *Mariner 10*, it will be an orbiter only.

Mercury has a diameter of 3,030 miles (4,880 km), so it is not a great deal larger than the moon, though it is denser and considerably more massive. Its weak gravitational pull means that it cannot have an Earth-type atmosphere, but there is an appreciating magnetic field so there must be an iron-rich core; indeed, this core used to be larger than the whole of the moon. The orbital period is 88 Earth days. It used to be thought that the axial rotation period was of the same length. If this had been so, Mercury would have kept the same face turned sunward all the time, giving permanent sunlight over one hemisphere, and permanent darkness over the other with only a narrow "twilight zone" in between. However, the real axial rotation period has been found to be 58.6 days so that conditions are not so extreme as this even though Mercury has a very peculiar calendar.

The orbit is appreciably eccentric; the distance from the sun ranges between 29 and 43 million miles (46 and 69 million kilometers). On the surface the interval between sunrise is 88 Earth days; there are two "hot poles" where the sun is overhead at perihelion and the temperature reaches 800°F (430°C). At night, a thermometer would register –300°F (–180°C). Under these conditions, it is hard to see how any form of life can exist on Mercury.

The first reasonably useful map of Mercury was drawn in the 1920s by the Greek astronomer E. M. Antoniadi. It did not show a great deal, but that was not Antoniadi's fault; he was the best observer of his time, and he was using the 33 in. (84 cm) refractor at Meudon (Paris)—but Mercury is never well placed, and never comes much nearer than 50 million miles (80 million kilometers) of us. We had to await results from the first space mission. *Mariner 10* was launched from Cape Canaveral on November 3, 1973. On February 5, 1974, it passed Venus at a range of 2,600 miles (4,184 km), and used the gravitational pull of that planet to put into an orbit which would take it to Mercury—the first example of what is termed "gravity assist" (some people have been known to refer to it as "interplanetary snooker"). *Mariner* made three active passes of Mercury; on March 29, 1974, September 21, 1974, and March 16, 1975. The last encounter was the closest, and *Mariner* flew by at only 203 miles (327 km) above the surface. By that time the instruments were starting to fail, and contact was finally lost on March 24, 1975. No doubt the spacecraft is still moving around the sun, and making regular passes of Mercury, but we have no hope of finding it again.

Like the moon, Mercury has highland areas, plains, mountains, valleys, and impact craters; the general aspect is decidedly lunar, but there are important differences in detail. Unfortunately, the same regions were available at each pass of *Mariner*, and only 45 percent of the surface could be imaged. There is no reason to think that the unexplored areas will be very different from those we have seen—but one never knows! The presence of a magnetic field was confirmed, and, as expected, the atmosphere was absolutely negligible.

The most imposing feature on Mercury is the Caloris Basin (Caloris Planitia) which contains one of the two hot poles—hence the name.

Opposite: Mercury, from *Mariner 10*. Part of the Caloris Basin is seen to the left, midway up the picture; unfortunately the other part has not been imaged, as the same areas of Mercury were shown during each of *Mariner 10's* active passes of the planet. Less than half the total surface was mapped.

(The hot poles are nowhere near the rotational poles; Mercury's axial inclination is less than 1 degree, so that it is "upright" with reference to its orbital plane.) Caloris is 810 miles (1,300 km) in diameter, and is bound by a ring of high, smooth mountain blocks, beyond which lies a second, weaker scarp. Its floor is crowded with detail—or rather, the part of the floor that we have seen; the other half was out of *Mariner*'s range. It was undoubtedly produced by a major impact, and its influence on the surface is widespread. Antipodal to it is a region of what is termed "weir terrain," with chaotic structures; it seems that shock waves from the Caloris impact traveled round the planet and then converged.

Almost half the surface imaged from *Mariner* is occupied by "intercrater plains," which are very ancient, and unlike anything found on the moon. There are also lobate scarps, high cliffs sometimes as much as 300 miles (500 km) long, which cut through other features and distort older landforms. Again there is no lunar counterpart.

Apart from Caloris, the largest circular structure is Beethoven, 400 miles (640 km) in diameter. There are many others well over 100 miles (161 km) across, and there are ray-craters, hills, and ridges.

A new Mercury probe, *Messenger*, was launched from Canaveral on August 3, 2004. After a decidedly tortuous journey it will be put into an orbit round Mercury in March 2011. Other missions are also planned, notably BepiColombo, a joint venture between Japan and the European Space Agency. No doubt these probes will add much to what *Mariner 10* told us, but manned expeditions to either Venus or Mercury seem to be a long way in the future. Fascinating though they are, these inner worlds will not give us a friendly welcome.

Left: The highlands of Mercury from *Mariner 10*. Superficially, the surface looks very much like that of the moon, although there are significant differences in detail. Active vulcanism ended a very long time ago, and today the surface is changeless—apart, possibly, from the results of occasional meteoroid impacts.

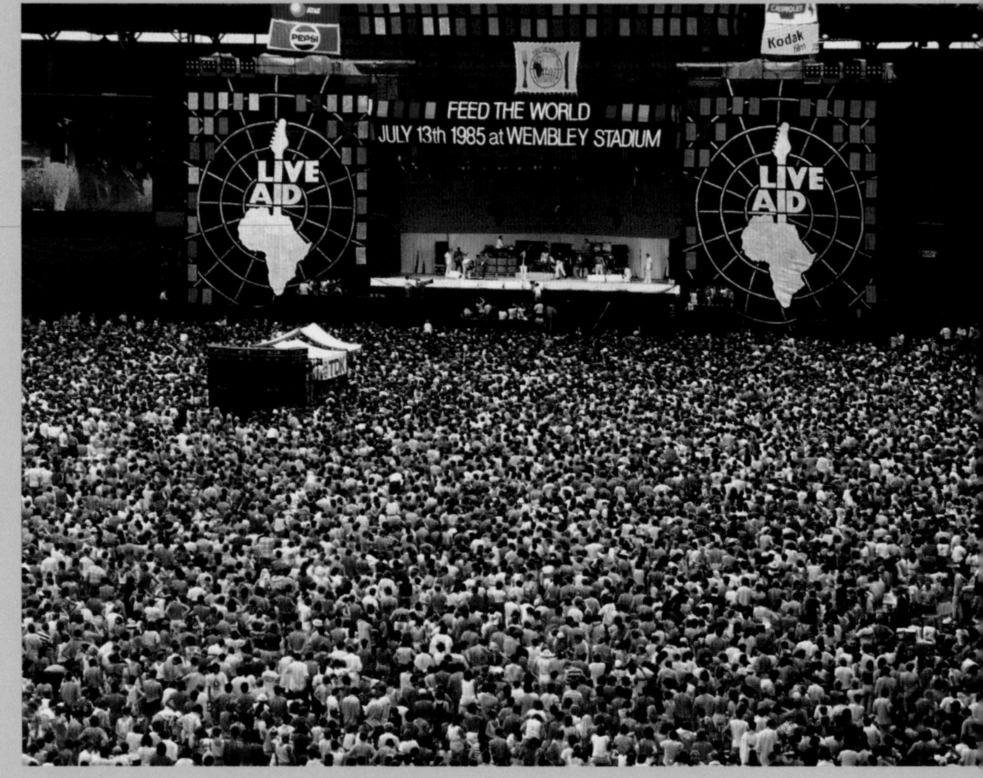

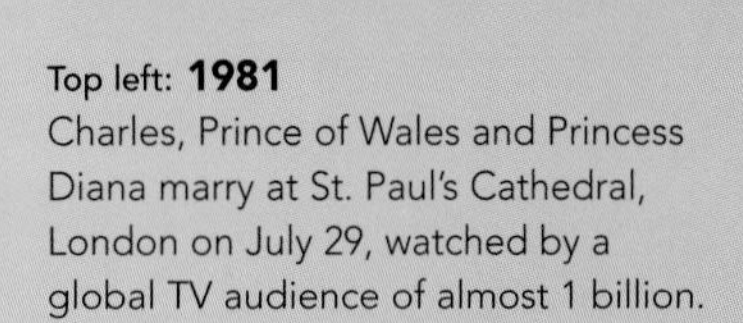

Top left: **1981**
Charles, Prince of Wales and Princess Diana marry at St. Paul's Cathedral, London on July 29, watched by a global TV audience of almost 1 billion.

Top right: **1983**
In Central Park people gather to hold lighted candles in memory of the death of an AIDS victim in New York. The vigil, attended by more than 1,500 people, symbolizes a growing public awareness of the disease.

Bottom left: **1984**
Steve Jobs unveils the new Apple Macintosh computer on January 30, 1984, in New York City. Jobs, with his business partner Steve Wozniak, had built the Apple computer in 1976 as a user-friendly alternative to IBM's personal computer.

Bottom right: **1985**
July 13 – Huge crowds gather in Wembley Stadium for the Live Aid concert. Watched by millions of people around the world on television, the concert raises huge amounts of donated money to help relieve a severe famine in Ethiopia.

1980–1989

Top left: **1986**
May 1 – The explosion at Chernobyl in Russia on April 26 causes a massive leak of radioactive gas into the environment, with devastating consequences. In the wake of the nuclear disaster, heavy equipment surrounds the nuclear plant, as safety work is carried out on Block number 4.

Top right: **1987**
U.S. President Ronald Reagan and Soviet leader Mikhail Gorbachev sign the Intermediate-Range Nuclear Forces (INF) agreement on December 8. By agreeing to eliminate medium-range missiles in Europe, the ceremony represents an important milestone in the cessation of the Cold War.

Above: **1989**
Crowds of West Berliners gather in front of the Berlin Wall, early on November 11. East German border guards demolish a section of the wall, creating a new crossing point between East and West Berlin—symbolically marking the end of the Cold War.

Opposite: Mars as imaged by *Viking 2* during its journey to the planet. The most prominent feature is the Valles Marineris, which is over 2,700 miles (4,345 km) long with a maximum width of 450 miles (724 km); the deepest part of the floor is 4 miles (6.4 km) below the crest of the rim.

06 THE EXPLORATION OF MARS

FACT AND FICTION

Venus may be like Earth in size and mass, but it is Mars that has always been regarded as the more likely abode of life. The Red Planet is much smaller than our world, with a diameter of only 4,200 miles (6,800 km); it is farther from the sun, and its atmosphere is painfully thin. Yet in many ways, it seems reasonable enough; it is much colder than the earth, but it is not in a condition of permanent deep freeze, and there are no obvious dangers, as there are with Venus. Astronauts going there will not have to contend with clouds of sulphuric acid, or temperatures high enough to melt lead.

Mars takes 687 Earth days to complete one journey around the sun. This is equivalent to 669 Mars days, or "sols," because Mars has a rotation period of 24 hours, 37 minutes. The axial inclination is almost the same as ours, so that the seasons are of the same type apart from being much longer. However, there is one important factor. The Martian orbit is appreciably eccentric, and the distance from the sun ranges between 124 and 154 million miles (200 and 250 million km). As with Earth, summer in the southern hemisphere falls at perihelion so that the southern summers are shorter and hotter than those in the north, while the winters are longer and colder.

Mars comes to opposition at a mean interval of 780 days—that is to say, in every other year. Not all oppositions are equally favorable; at its best the magnitude may reach –2.8, brighter than any other planet apart from Venus, while at its faintest, Mars is not much brighter than the Pole Star, though its redness makes it easy to identify. It is this hue which led to Mars being named after the God of War; red is the color of blood.

When Mars is well placed, even a small telescope will show surface features, because the atmosphere is (usually) transparent. There are white polar caps which wax and wane with the Martian seasons, there are wide reddish-ochre areas, and there are dark markings which alter little in size and shape. The first reasonably good map was drawn by two German astronomers, Beer and Mädler, in the 1830s. Others followed, and names were

given to the various surface features. It seemed natural to believe that the red areas were land masses, and that the dark patches were seas, while the polar caps were assumed to be ice. There seemed no reason to doubt that life existed there—and why should there not be intelligent "Martians"?

There were interesting developments in 1877, during a very favorable opposition, and Mars was as close to Earth as it ever can be—just under 35 million miles (56 million km) away. Observations were made by the Italian astronomer G. V. Schiaparelli, from Milan; he used an excellent 9-inch (23-cm) telescope, and drew up a new map of the planet, giving the surface features the names we still use, though in modified form. He showed the ice caps, the red "deserts," and the dark "seas," but he also showed strange lines which he called "canali." This is the Italian for "channels," but inevitably it was translated as "canals," and equally inevitably it was suggested that we might be seeing artificial waterways, built by Martian engineers to make a planet-wide irrigation system. Schiaparelli himself was careful to keep an open mind, but others were less cautious. Percival Lowell, who set up an observatory at Flagstaff in Arizona specially to observe Mars, and equipped it with a fine 33-inch (84-cm) refractor, even said "that Mars is inhabited by beings of some kind or other is as certain as it is uncertain what those beings may be."

Lowell's drawings were certainly peculiar. The canals even showed a remarkable ability to become double—if one channel could not cope with the flow of water a second could presumably be opened! And in the following years other observers started to record canals; some of the drawings were made with small telescopes. But there were sceptics too. Some well-equipped astronomers could not see the canals at all, or else recorded them as disconnected patches. By around 1930 the idea of intelligent Martians had been generally abandoned, but the canals lingered on into the space age. Disappointingly, we now know that they were merely optical illusions. When looking for details at the limit of visibility, it is only too easy to "see" what you half expect to see. (I have made many drawings of Mars with Lowell's telescope. Canals are conspicuous only by their absence.)

Our ideas about Mars have changed dramatically over the past half-century. It is worth looking back to 1957, the start of the space age, and see where we were right and where we were wrong.

1957: Mars is a cold world, but at daytime at the equator, in summer, the temperature can rise well above freezing.

2007: RIGHT. A summer day at the equator will rise to a balmy 40°F (4°C), although by nightfall it will become bitterly cold.

1957: The atmosphere is made up chiefly of nitrogen, with a ground pressure of about 85 millibars.

2007: WRONG. The main constituent is carbon dioxide, and the pressure everywhere is below 10 millibars.

1957: The red areas are deserts, similar to our Sahara.

2007: WRONG. The red areas are coated with reddish oxides—"rust."

1957: The polar caps are due to a layer of frost, no more than a few centimetres thick.

2007: WRONG. The polar caps are very thick indeed.

1957: The dark areas are swampy areas, and probably old sea beds.

2007: WRONG. Some are high plateaux, and they are dry.

1957: The dark areas are covered with vegetation.

2007: WRONG. There is no vegetation on the Martian surface.

1957: The polar ices melt with the arrival of warmer weather in spring and summer.

2007: WRONG. The ice sublimates (changes directly from solid to vapor).

1957: The axial inclination is almost the same as that of the arth.

2007: RIGHT. The axial inclination is 24 degrees against our 23 degrees.

1957: When a cap shrinks, and water vapour is released, a "wave of darkening" of the plant-covered areas sweeps equatorward, because the plants revive when moisture reaches them.

2007: WRONG. There is no "wave of darkening." There are no plants.

1957: Liquid water can exist on the surface; there may be ponds or lakes—even shallow seas.

2007: WRONG. The low atmospheric pressure means that the liquid water would immediately evaporate.

1957: The surface is gently undulating, with no high mountains, deep valleys, or large craters.

2007: WRONG. There are high peaks, deep valleys, and many very large impact craters.

1957: There is no magnetic field.

2007: WRONG. There is a magnetic field but it is extremely weak.

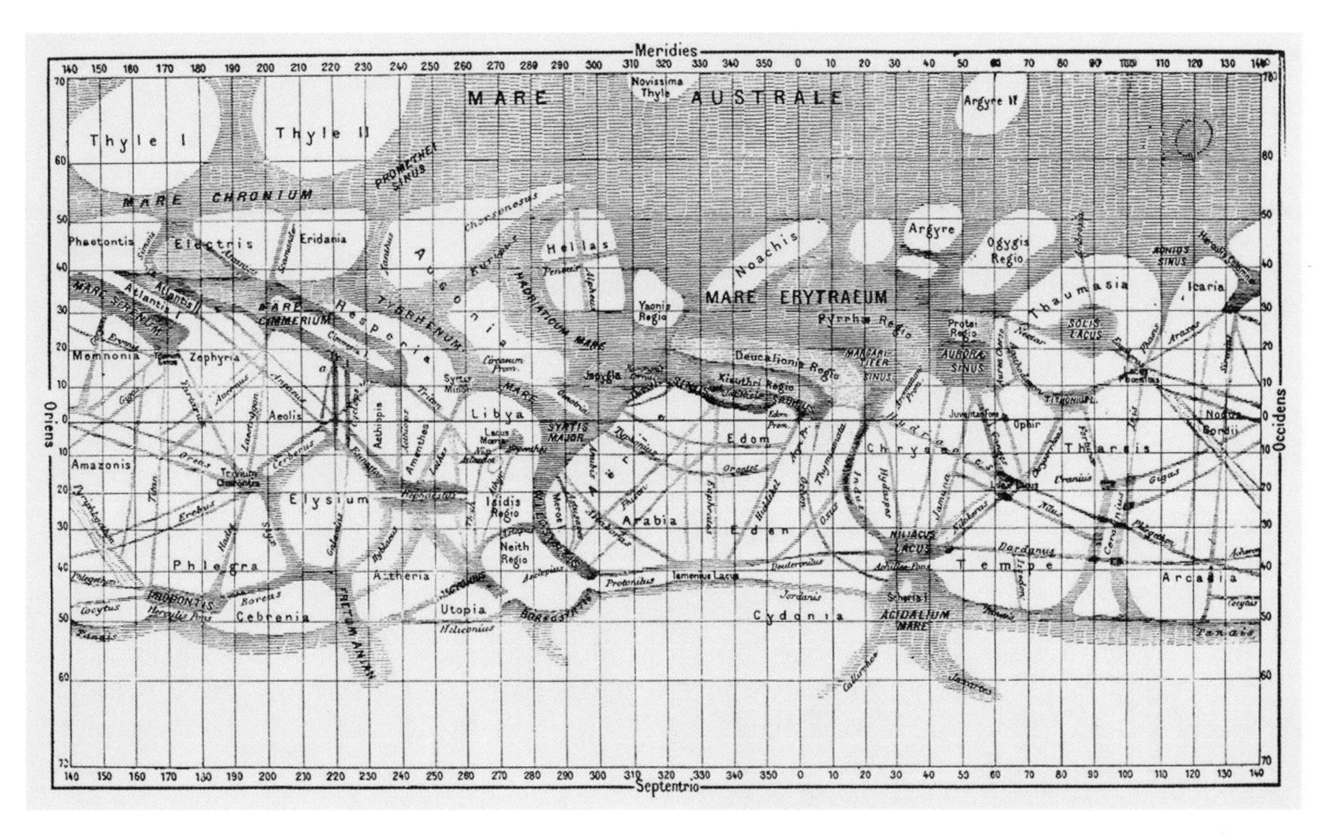

Above: Giovanni Schiaparelli's map of Mars from 1888, showing some of the "channels" that started the "canals" debate.

Right: Mars as seen by Percival Lowell in 1907. Unlike Schiaparelli, Lowell confidently identified a "canal network," but in fact the canals do not exist; they were an optical illusion.

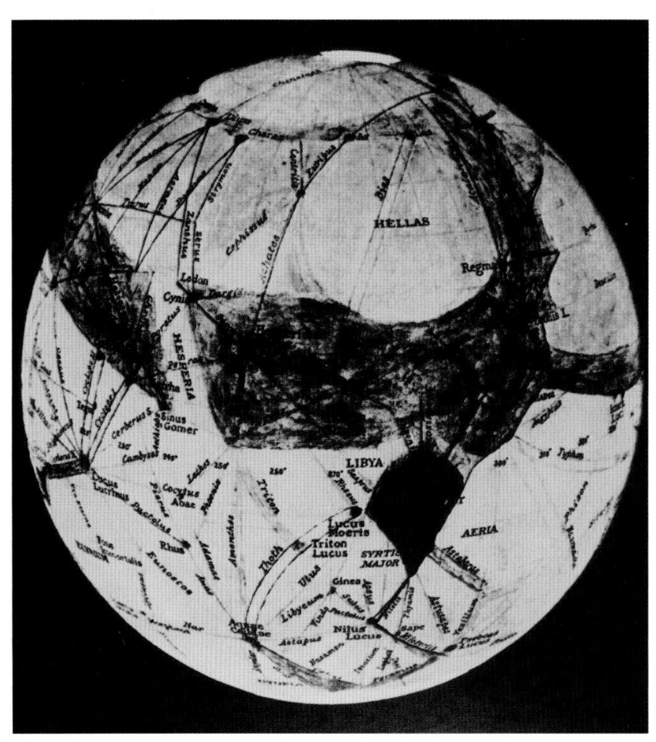

Obviously there has been a great deal of progress. Another idea, current for some time, was that rather than being made of water ice, the polar caps consisted of solid carbon dioxide—"dry ice," of the type found in an old fashioned ice cream seller's barrow. But the flight of *Mariner 4* brought about a complete change of thought.

FIRST VISITS TO MARS

Mariner 4 was not the first Mars probe. As usual, the Russians took the lead, and sent up *Mars 1* on November 1, 1962. It was put into the right path, but contact with it was lost when it had receded to a distance of 66,000 miles (106,000 km), and nothing more was ever heard from it. In fact, up to the time of writing (2007) the Russians have had absolutely no success with Mars—which is surprising in view of the fact that they have had such good results with Venus, which would seem to be a much more difficult target. The Japanese fared no better with their one mission, *Nozomi* (1998). Europe's *Mars Express* (2003) is operating well, but otherwise all our data came from the probes launched from the United States.

Nevertheless, NASA's program had an inauspicious beginning. *Mariner 3* (November 1964) was a prompt failure, and sped off into space uncontrollable and uncontactable (presumably it is still going around the sun). However, its twin, *Mariner 4*, more than compensated for this setback. It began its journey from Cape Canaveral on November 28, 1964, and flew past Mars, at a range of 6,084 miles (9,791 km), on the following July 15. It sent back 21 images, of which several were excellent. It then went on its way; contact with it was maintained until December 20, 1967.

There were three results of special importance. Firstly, goodbye to the canals. Secondly, the atmosphere was much thinner than had been expected. If the previously favored grand pressure of 85 millibars had been correct, the air at the surface would have been about as dense as the earth's air at 52,000 feet (16,000 m) above sea level; we could not breathe it. But *Mariner* showed that the ground pressure is about the same as that of our air at around 120,000 feet (90,000 m), which ruled out any advanced life forms. Moreover, the atmosphere was over 95 percent carbon dioxide, with nitrogen accounting for less than 3 percent. Thirdly, the images showed craters apparently similar in type to those of the moon. As a world, Mars seemed to be more lunar than terrestrial.

Mariners 6 and *7* followed in 1969. They were fly-by missions, and were successful, sending back more images as well as confirming the earlier results. Craters, no doubt of impact origin, were plentiful; one interesting picture was obtained of Hellas, a bright feature in the southern hemisphere, which had been regarded as a snow covered plateau, but which turned out to be a deep basin often filled with clouds. Unwary observers have often mistaken it for an extra polar cap. (It adjoins the Syrtis Major, the darkest feature on Mars, which is itself a plateau with sloping sides. Early observers had called it "Lookyer Land," after a famous British astronomer of the time; Schiaparelli renamed it.) Yet after these first missions, many astronomers felt disappointed. Mars gave the impression of being nothing more than a larger version of the cratered, inert moon.

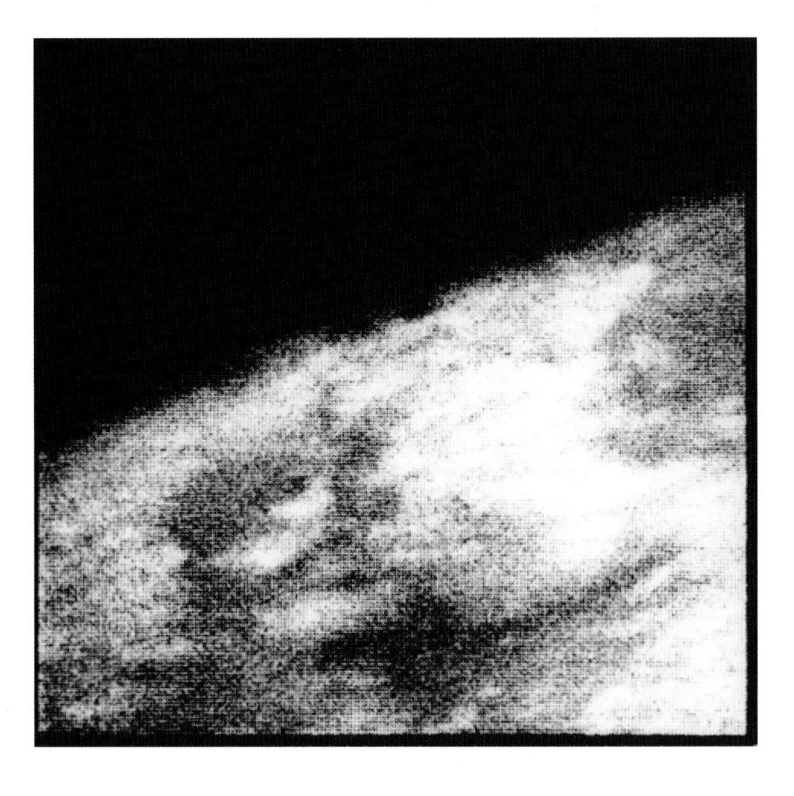

Opposite: A NASA publicity photograph, showing one of the *Mariner* craft during construction. It was a very complex vehicle, and the greatest care had to be taken during the assembly; there were also many tests to be carried out—for instance, the vehicle had to be sterilized to make sure that no contamination was carried from the earth to Mars.

Above: The first image of Mars sent back by *Mariner 4*. It shows an area over 180 miles (290 km) across, but without much detail. *Mariner 4* was a fly-by probe; it made its closest approach to Mars on July 14, 1965, and returned 21 images. Contact was lost on December 21, 1967; presumably the probe is now in solar orbit.

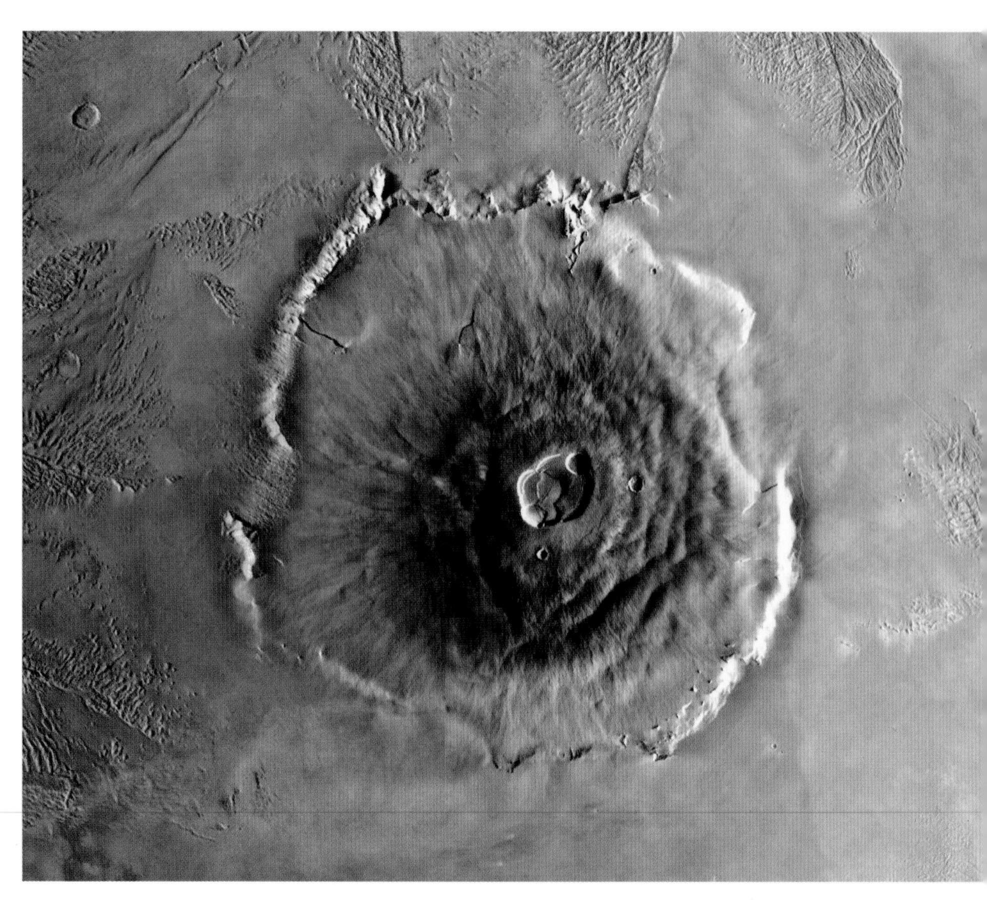

We now know that, by sheer bad luck, *Mariners 4*, *6*, and *7* surveyed the least interesting regions of Mars. Further surprises were in store. *Mariner 8* (May 9, 1971) failed because the second stage of the rocket launcher did not ignite, but *Mariner 9* (May 30, 1971) reached Mars on November 13 and entered orbit around the planet. By the time it lost contact, on October 27, 1972, it had returned over 7,000 images, and for the first time, we saw the gorges, the dry riverbeds and, most impressive of all, the giant volcanoes. The loftiest of these, Olympus Mons (Mount Olympus) rises to over 15 miles (24 km), three times the height of Everest, and is crowned by a complex 40 miles (65 km) caldera. It is thought to have been active less than 300,000 years ago, and it may not be quite dead, even today, although major eruptions on Mars no longer occur.

Incidentally, altitudes on Mars cannot be referred to sea level, as ours are, because there are no seas on Mars. Instead, we use a datum line where the atmospheric pressure is 6.2 millibars. The two hemispheres are different; generally speaking the southern part of the planet is thickly cratered, much of it lying between 1.5 and 2 miles (2.5 and 3 km) above the datum line, while the northern hemisphere is less cratered and lies below the datum line. However, the two largest and deepest basins, Hellas and Argyre, lie in the south. The main volcanic area is the Tharsis Ridge; here we find the four tallest volcanoes, Olympus Mons, Ascraeus Mons, Pavonis Mons, and Arsia Mons. The Tharsis Ridge straddles the equator, but only Arsia Mons is in the southern hemisphere.

Dust storms are common on Mars, and may be global. One such storm was in progress when *Mariner 9* arrived, and it was several weeks before the dust cleared away. For a while, Mars seemed almost featureless, and *Mariner*'s cameras could not see through to the surface. Low clouds are very frequent, and resemble our cirrus, but for most of the time the Martian atmosphere is more or less transparent. Whether it will prove to be an adequate screen against harmful incoming radiation is still a matter for debate.

The greatest canyon system on Mars has been named the Valles Marineris in honor of the pioneer spacecraft. It extends for a total length of over 2,700 miles (4,300 km), with a maximum width of 370 miles (600 km) and a greatest depth of more than 4 miles (6 km); it certainly dwarfs the earth's Grand Canyon. It begins in an amazingly complex canyon region known officially as Noctis Labyrinthus, but often nicknamed "the Chandelier" because of its outward appearance; it was

Above: Part of the Meridiani Terra (formerly called Meridiani Sinus), imaged by *Mariner 6* in July 1969 from a range of about 2,100 miles (3,380 km). Craters are shown, but there were no great volcanoes in the areas imaged by either *Mariner 6* or its twin, *Mariner 7*. *Mariner 6* returned 76 images, and flew over the Martian equator. It is still orbiting the sun.

Above right: Olympus Mons, from *Mariner 9*. This is the greatest known volcano in the solar system, three times as high as our Everest and crowned by a 40-mile (64 km) caldera. The slopes are fairly gentle; lava flows are prominent, and there is an extensive "aureole" of blocks and ridges around the base.

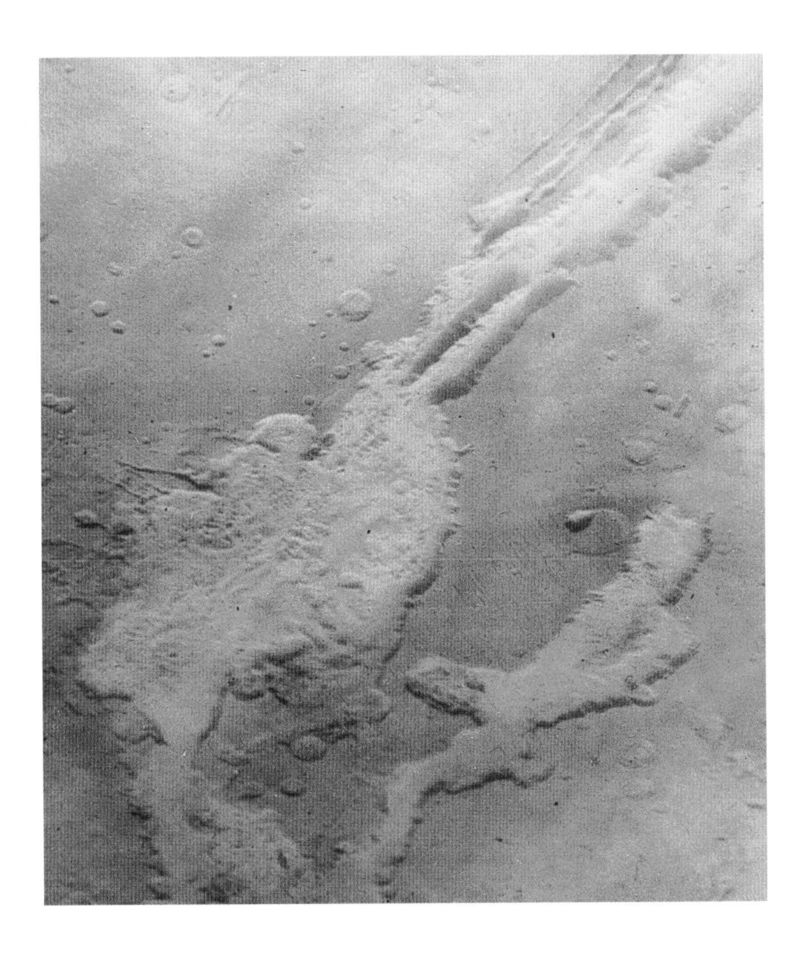

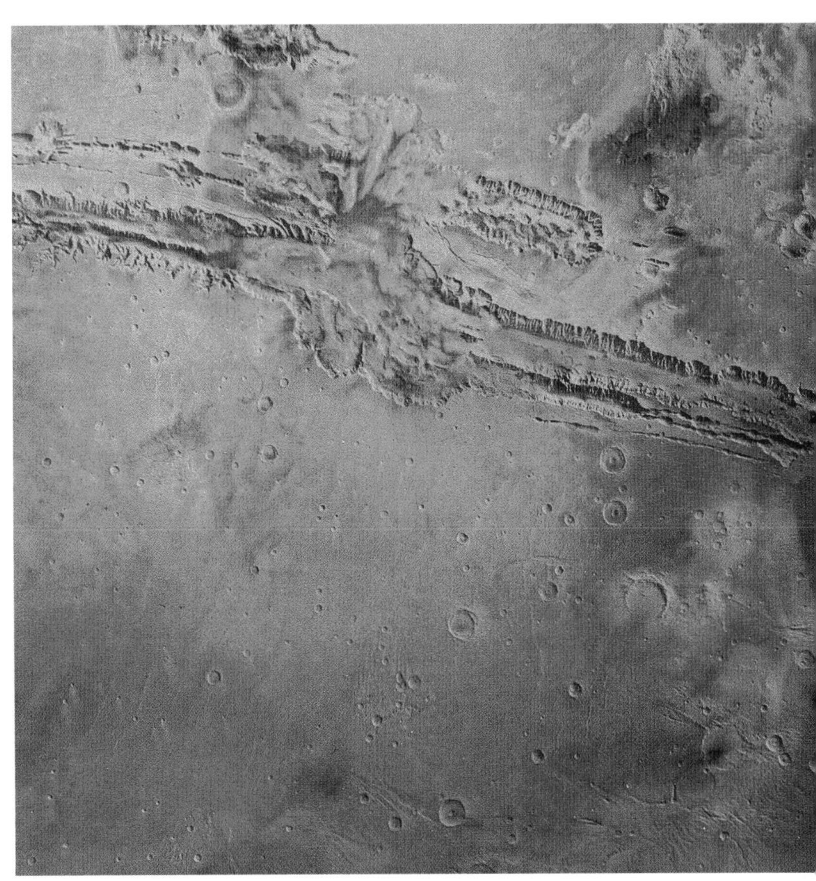

once called Nocis Lacus—Lake of the Night—when the features were believed to be waterways. From the Chandelier it runs eastward, ending in blocky terrain close to another prominent dark feature—Margaritifer Terra, formerly Margaritifer Sinus, the Gulf of Pearls. Inappropriate though they have proved to be, these old names are certainly attractive.

There are features that look so like dry riverbeds that they can hardly be anything else. Some are hundreds of kilometres long. Several channels debouch into the plain which is known as Chryse—and this was one reason why Chryse was chosen to be the site of NASA's first soft-lander, *Viking 1*, in 1976.

There were two *Vikings*—identical twins, designed to carry out the same programs, but in different areas. *Viking 1* was to be aimed at Chryse, and *Viking 2* at the more northerly plain of Utopia. Each consisted of an orbiter and a lander; after the lander had reached the Martian surface the orbiter would go on circling the planet, acting as a relay and also continuing the surveys made by the earlier spacecraft. The landing procedure was carefully worked out. The lander would separate from the orbiter and plunge into the atmosphere; its speed would be reduced by using retro rockets and parachutes (thin though it is, the atmosphere is substantial enough to make parachutes useful). The final touchdown speed would be no more than 6 mph (10km/h). There were obvious dangers; if the lander had come down on a large rock, or fallen in a deep crevice, the result would have been disastrous, and the investigators privately admitted that they would be satisfied if just one of the landers survived. In fact both operated excellently, and by the end of 1976 many of Mars' cherished secrets had been solved.

The first image sent back by *Viking 1* immediately after landing showed a rusty-red, rock-strewn landscape; the rocks were clearly of volcanic origin, and there was abundant evidence of windblown material. Chryse was a chilly place; temperatures ranged between –120°F (–84°C) at dawn to –22°F (–30°C) in early afternoon. One major surprise was the color of the daytime sky; instead of being dark blue, as many people had expected, it turned out to be yellowish-pink. All in all, it may be said that *Viking 1* was a triumph. So was *Viking 2*; the scene at Utopia was similar to that at Chryse, although there were differences in detail.

Searches for life were undertaken. Each *Viking* stretched out a "grab," collected material and drew it back into what was to all intents and purposes a tiny but superbly efficient laboratory set up inside the spacecraft. The results of the various tests were then transmitted to Earth. There had been real hope of finding microscopic life forms,

Above and above right: Two views of the Valles Marineris, imaged by *Viking 2*. It is apparently a tectonic feature rather than having been cut by flowing liquid; it makes the earth's Grand Canyon seem puny!

possibly similar to those which manage to exist in Death Valley or the Dry Valleys of Antarctica, but although the results of the analyses were in some ways puzzling there were no definite signs of organic material. Investigators began to think that Mars was totally sterile after all.

There was a flurry of excitement in 1993 when a meteorite found in Antarctica and cataloged as ALH 84001 was analyzed and found to contain very small features which might be primitive organisms. The significant point was that the meteorite was widely believed to have come from Mars, blasted away from the surface by a powerful impact and put into a path which led it on to collide with Earth. ALH 84001 and other meteorites of the same type may indeed be Martian, but even if the features do turn out to be primitive life-forms they are very probably due to Earth contamination. After all, the meteorite has been lying in Antarctica for a very long time.

RECENT MISSIONS

The next successful lander was *Pathfinder*, which was launched from Canaveral on December 4, 1996. (In the interim there had been two spectacular failures, NASA's *Mars Observer*, which lost contact for reasons unknown, and Russia's massive *Mars Observer*, which fell forlornly into the sea almost immediately after blast-off.) *Pathfinder* reached Mars on July 4, 1997, and came down in the region of Ares Vallis, almost 20 degrees north of the equator. This time there was no attempt to make a gentle landing. *Pathfinder* was encased in airbags, and impacted at a speed of 55 mph (89 km/h). It then bounced—not once, but at least fifteen times; the first bounce took it up to well over 500 feet (150 m) above the surface. It ended in an upright position, and on Sol 2—the second day on Mars—a hatch opened, and out crawled a tiny wheeled rover, *Sojourner*, about the size of a household television. It made its way down a ramp, which was already in position, and emerged on to the surface, ready to begin its travels.

Sojourner was not a quick mover; its top speed was less than ½ in. (1.3 cm) per second, but it carried an amazing collection of instruments. For example, it could make chemical analyses of rocks that were encountered. It could not be expected to travel very far, but it could be guided by the operators at Mission Control, so many millions of kilometers away, and could therefore avoid obvious obstacles. Data

Above: View from the camera of *Viking 1* (1976), which touched down in Chryse, the "Golden Plain," fortunately avoiding rocks littering the landing site. This was actually the first picture to be received direct from the Martian surface—(latitude 22.4 degrees N, longitude 47.5 degrees W). The image quality was better than the NASA planners had dared to hope.

Opposite: *Pathfinder* on Mars, July 1997. The spacecraft came down in Ares Vallis. It carried *Sojourner*, the first Mars rover, which moved around making analyses of the rocks. It could be controlled from Earth, and sent back 550 images before contact was finally lost in October. Undoubtedly the area was once covered with water.

were sent back using the main lander as a relay. By the time it lost contact on October 8, 1997, it had sent back 550 images, while *Pathfinder* itself (named by NASA "the Sagan Memorial Station" in honor of the astronomer Carl Sagan) transmitted 16,000.

The landing site had been chosen with the utmost care. Ares Vallis is a long valley almost certainly scoured out by water. It begins in the hilly region of Margaritifer Terra, runs through the highlands of Xanthe Terra and ends in the "Golden Plain" of Chryse; the actual landing site was an ancient flood plain, 530 miles (850 km) southeast of the touchdown point of *Viking 1*. Ares Vallis itself had once been a raging torrent, and it was thought that rocks of many different types would have been swept down on to the floodplain; presumably the area had once been a lake. In the event, NASA's choice proved to be well justified.

The camera on the Sagan station showed a landscape with dunes, mounds, and rocks of all kinds; in the distance there were two elevations christened the Twin Peaks, although they were not actually very high. The maximum daytime temperature was –14°F (–10°C), going down to below –100°F (–73°C) at night; the winds reached 20 mph (32 km/h), though in that thin atmosphere they had little force, and the air pressure was under 7 millibars, too low for liquid water to persist.

There were dust devils, and crystal clouds around 10 miles (16 km) up. But, of course, the main results came from *Sojourner*, which was able to carry out chemical analyses of the rocks—which were given graphic unofficial names such as Sausage, Boo Boo, Desert Princess, Mermaid, and Barnacle Bill! All were volcanic, and there was layering and bedding, suggesting a sedimentary origin. Evidence was mounting that the landing area had once been covered with water.

Three failures followed. Japan's *Nozomi* probe (launched on July 3, 1998) developed so many faults that any hope of reaching Mars had to be abandoned. *Mars Climate Orbiter* (launched from Canaveral on December 11, 1998) was lost because of an incredible blunder; it was ready to go into Mars orbit when it was sent instructions not in metric, but in imperial units, which it did not understand. It dived into the Martian atmosphere, and burned up. NASA's next probe, *Mars Polar Lander*, reached the planet on December 3, 1999, but nothing more was heard from either it or the two smaller landers that it carried. All this was decidedly depressing, but the twenty-first century started on a happier note, with two successful orbiters, NASA's *Odyssey* (reached Mars on October 24, 2001) and the European Space Agency's *Mars Express* (December 26, 2003). Sadly, Europe's first attempt at a landing

failed; *Beagle 2*, transported by *Mars Express* and separated six days before insertion, may well have landed intact but sent back no signals after arrival. Perhaps it—or what remains of it—will be located one day.

All previous efforts were surpassed by the two Mars Exploration rovers, *Spirit* and *Opportunity*, which were launched in mid-2003 and came down gently on Mars in January 2004. They were scheduled to remain active for a mere 90 days, but were still moving happily around in the summer of 2006—well over one Martian year later. Their results exceeded all expectations.

The rovers were identical twins. On entering the Martian atmosphere, a parachute was deployed to reduce speed. Retro rockets were fired just before impact, and airbags were inflated to cushion the landing. Once down, the spacecraft bounced several times before coming to rest. The airbags were then deflated and retracted, and the rover itself was able to drive off the lander, and work could begin. The rovers were not large; each was 5.2 feet (1.5 m) long, and on Earth weighed 384 pounds (174 kg), reduced to a modest 144 pounds (65 kg) on Mars. Each was crammed with instruments; they could analyze the rocks and the surface materials, for example, and use an "abrasion tool" to scrape rocks and examine what lay just below the surface layer. Observations of all kinds could be carried out, and, of course, photography was a major concern. Each rover had six wheels, and a top speed could hurtle along at a rate of 12 feet (3.7 m) per hour. Solar power was used together with rechargeable batteries. The rovers were able to communicate with the orbiters, *Mars Express* and *Odyssey*. They were certainly versatile.

Spirit, first to land, came down in the 103 mile (166 km) crater Gusev, thought to have once been a lake because a huge channel system over 500 miles (800 km) long, Ma'adim Vallis, drains into it. Inside it is a 450 feet (137 m) crater, Bonneville, visited by *Spirit* during its first foray. About 1.5 miles (2.5 km) from the landing point lies a range of hills, the Columbia Hills. On August 21, 2005, *Spirit* managed to climb to the top of the highest of these, Husband Hill, and had a magnificent panoramic view of the scene below. (All these names are as yet provisional, though in due course they will no doubt be ratified by the International Astronomical Union.) Opportunity touched down three weeks later on the opposite side of Mars; the site was the plain known as Meridiani Planum, and fortuitously the spacecraft plumped into a tiny crater, Eagle. The area was chosen because it had been found to be rich in hematite, a mineral made up of iron and oxygen, which is often formed in liquid water. *Opportunity* did indeed find clear signs of the past presence of flowing water, and probably the site lay near the shoreline of a salty sea which had dried up long ago. A rocky outcrop, El Capitan, was found to contain tiny hematite-lined spheres nicknamed "blueberries," and there were deposits of a mineral called jarosite, which tends to be formed in groundwater.

Over a Martian year after arrival, *Spirit* and *Opportunity* had each covered almost 4 miles (6.5 km), and had sent back vast quantities of data. Inevitably there had been problems; in late 2004 one of *Spirit*'s wheels gave trouble, overcome by driving in reverse, while on two occasions in 2006, *Opportunity* became stuck and had considerable difficulty in freeing itself. But all in all, the missions could hardly have gone better.

There is now no reasonable doubt that water was once plentiful on Mars, and presumably conditions there were once suitable for life. Yet we do not know whether life did appear—and we do not know whether life forms still exist. If they do, they must be primitive and probably below the planet's surface; when the temperature was higher and seas existed, it seems unlikely that life could have evolved very far before conditions deteriorated. Future explorers are not hopeful of finding fossils of Martian dinosaurs.

When will men reach Mars? It is hard to say; there are so many factors to be taken into account, political as well as scientific. It may well be that the first piloted missions will use one of the two midget satellites, Phobos and Deimos, as a convenient natural space station; both have been surveyed by the various probes, and have been irregularly-shaped bodies; Phobos has the longest diameter of less than 20 miles (32 km). Deimos has less than 10. No doubt both are captured asteroids rather than bona fide satellites.

NASA's *Mars Reconnaissance Orbiter* began circling the planet in March 2006, and at once started to send back images and data of the highest quality. Future missions have been planned for the near future; although we cannot yet claim to have a full understanding of Mars, we do at least know the types of problems that the pioneer colonist will face. The Red Planet awaits us.

Opposite: Victoria Crater, imaged from orbit. The *Opportunity* rover can be seen on the crater rim, at the 10 o'clock position. By this time, *Opportunity* had traveled 5 miles (8 km) from its landing point. Victoria Crater is 0.5 miles (0.8 km) in diameter and well formed; the scalloped shape of the rim is due to erosion and downward movement of the crater wall.

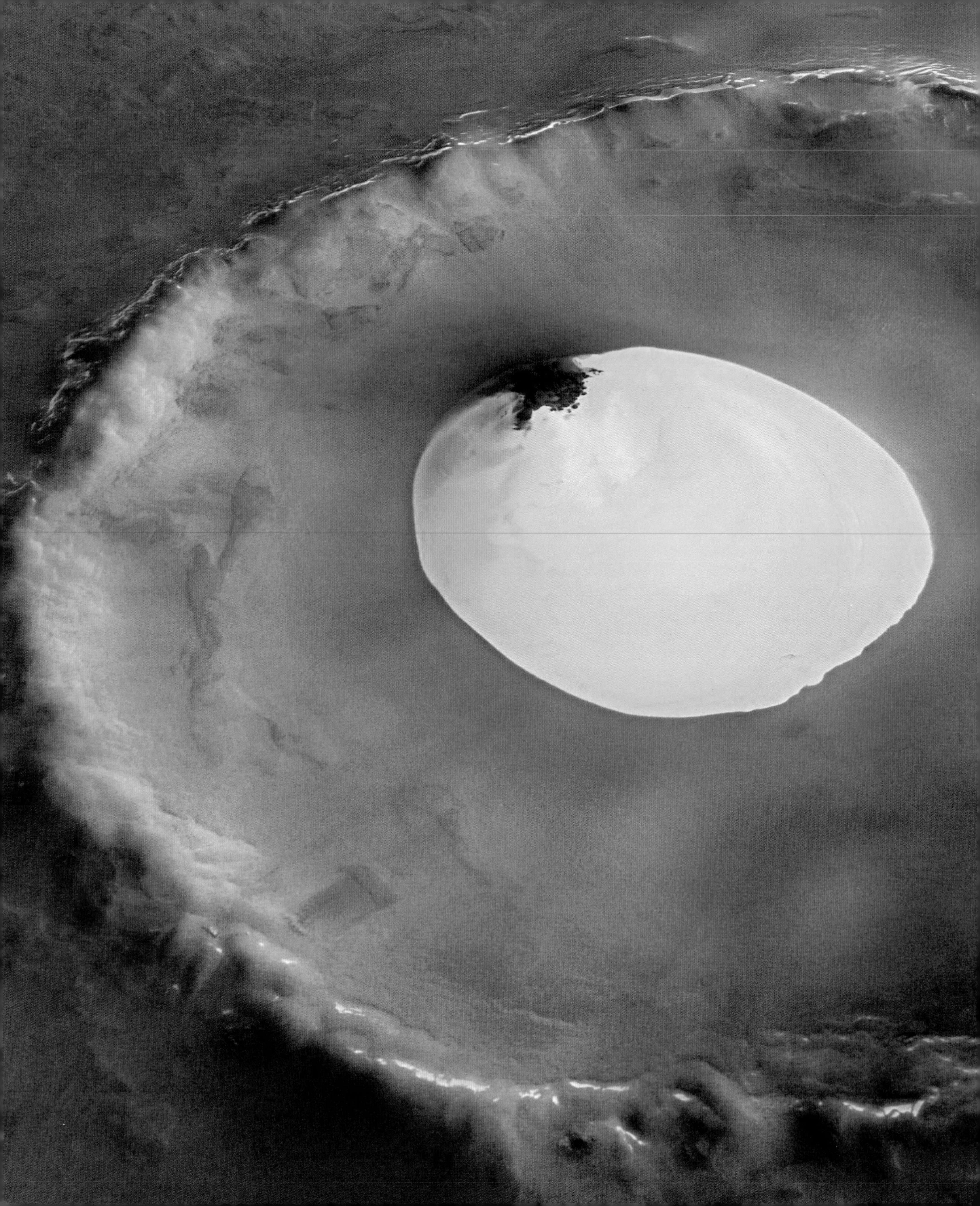

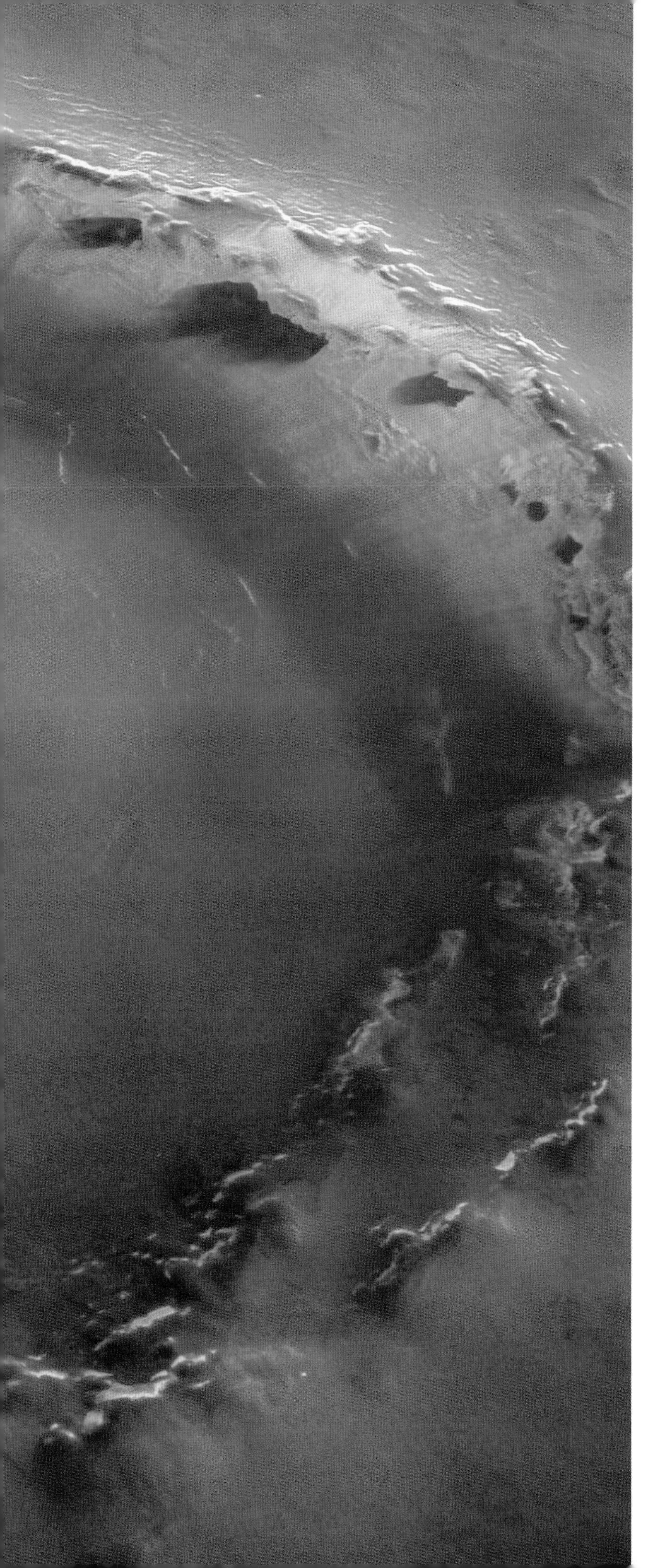

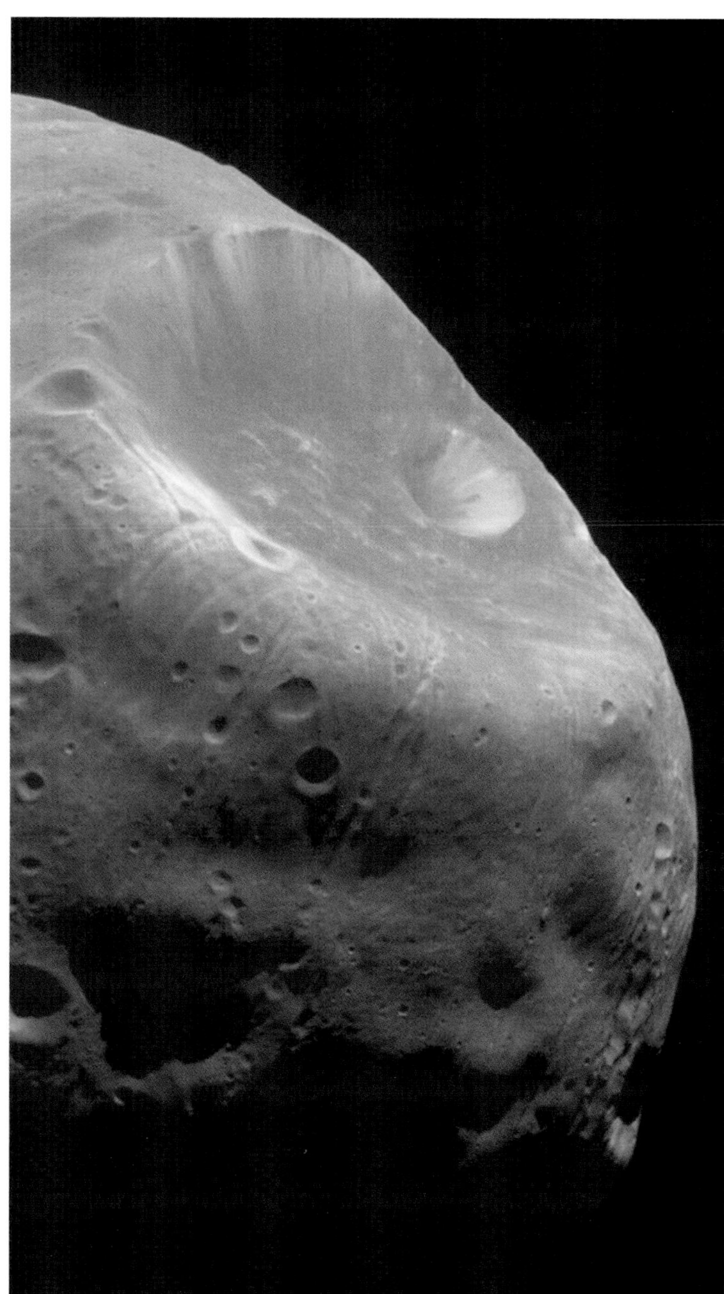

Above: A spectacular view of Phobos, one of the two Martian satellites, taken by *Mars Global Surveyor* in 1998. The large crater on the upper right, Stickney, is 6 miles (10 km) across, which is almost half of the width of Phobos. The satellite probably came close to breaking up when it was hit.

Left: A vast patch of water ice on the floor of a 22 mile (35.5 km) Martian crater; the crater is about 1.2 miles (1.9 km) deep. The picture was taken with the high-resolution stereo camera on ESA's *Mars Express* spacecraft; the colors are close to natural.

THE 50 GREATEST SPACE IMAGES

Trying to select fifty great images from the space era—and, in addition, endeavoring to place them in an order of precedence—is a daunting but fascinating task. It helps to have lived through all those years: as I write this, I glance at a book of cuttings and see that within a few days of Sputnik 1's launch on October 4, 1957, I contributed two pieces to the Financial Times on "Russia's Satellite Experts" and "Science in the USSR."

Millions of images have been returned from space. In having to select so few, one's inclination is to record the "firsts" along the way—to honor the major steps that got us to where we are today. The problem with that is that images of the first major events are often poor by today's standards and would fail to impress. Some had to be included, of course—a view of Sputnik 1 could not possibly be omitted from the fifty.

For the rest, a choice of themes seemed to be sensible, arranged broadly, although not exclusively, so in an approximate chronology to give a context of the progress that has taken place. The subjects chose themselves: rockets, machines in space, men in space—with *Apollo* prominent, interplanetary exploration and the earth from space, not forgetting to record the important applications story which is one of the greatest achievements of the first fifty years: Can we imagine a world now without communications and meteorological satellites?

The vast mass of space era images have been a means of supplying information—they have not been taken with the eyes of an artist or talented photographer. The films that came back from the moon between 1968 and 1972, for example, were largely filled with frames intended for review by geologists and other scientists. However, in those films and in the images from all other missions, on occasion there was evidence of a "seeing eye," which recognized a powerful subject and elsewhere the very subject matter itself yielded a beautiful, awe-inspiring or mystifying record which gave those of us who would never leave the earth's gravity field an idea of what "space" was like. In some of those images, reality could become fantasy and very occasionally one realized that chance or serendipity knows no orbital limits.

Having had the privilege of spending most of my working life being involved in some way with imaging in and from space, I ultimately decided to place my trust primarily in memory—to sit with a blank piece of paper and, starting from 1957, work through the themes and note memories of images which had impressed me so much that they were never to be forgotten. A review of various pertinent books followed but the changes were few. Responding to the challenge of grading the images was much more difficult. How does one compare a superb *Gemini* view of the earth from the 1960s with a *Cassini* view of Mimas and Saturn's rings taken in 2005? The grading must reveal more about my evaluation of the images and my interests than any absolute judgement on their importance in the history of spaceflight's first fifty years.

Other spaceflight professionals might include some of my chosen images in their list if asked to carry out the same task, but their selections would doubtless differ considerably. My hope is that, while for many readers there will doubtless be old favorites, some of the selection will be new and intriguing—and that taken together they will impress as a record of the first half century of a time when humankind broke free from the bounds of earth.

H. J. P. Arnold

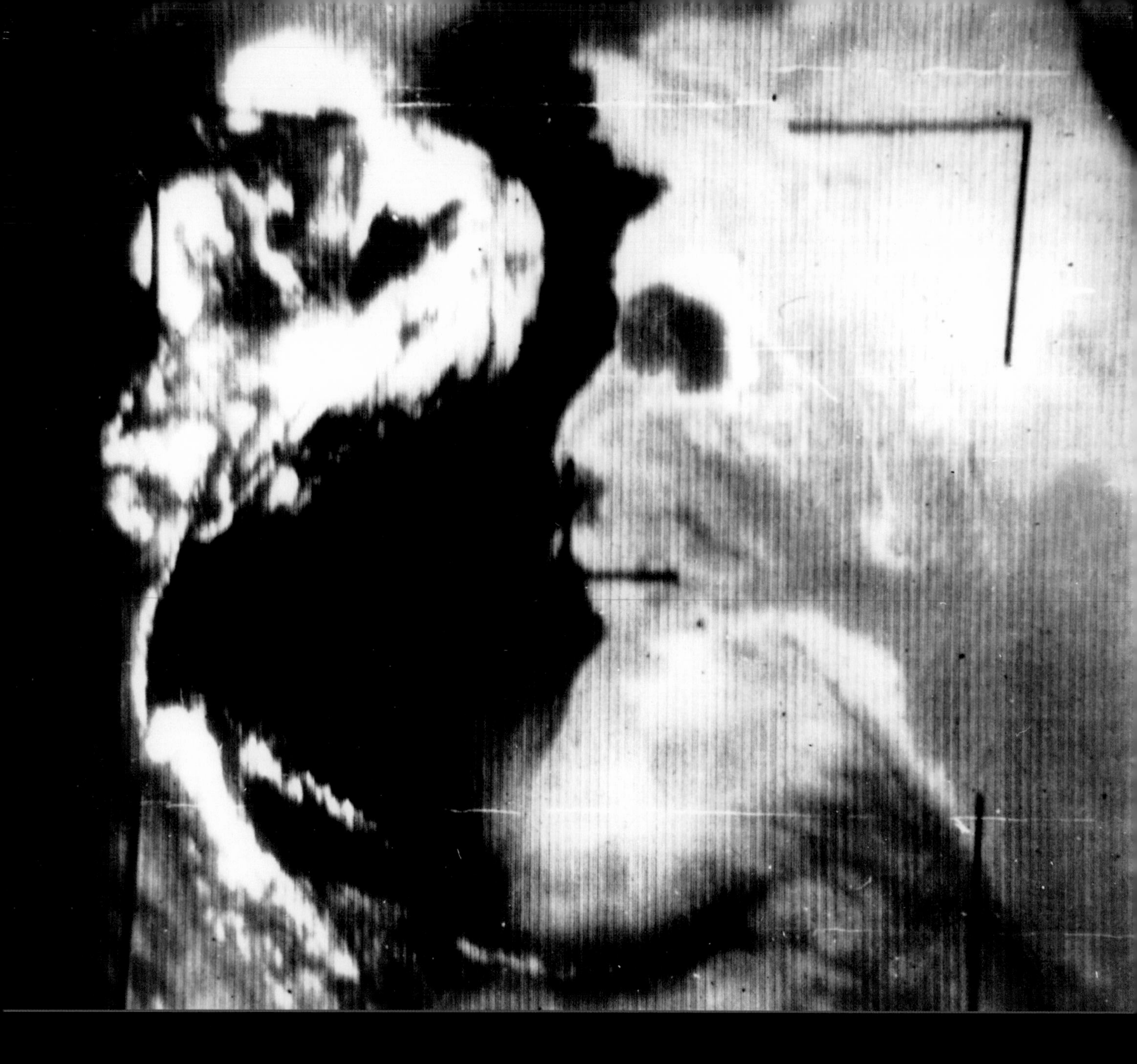

50

It has sometimes been alleged that governments have indulged in conspiracies to hide the truth about such subjects as visits to the earth by aliens. Perhaps the most extreme allegation has been the claim that the *Apollo* missions did not go

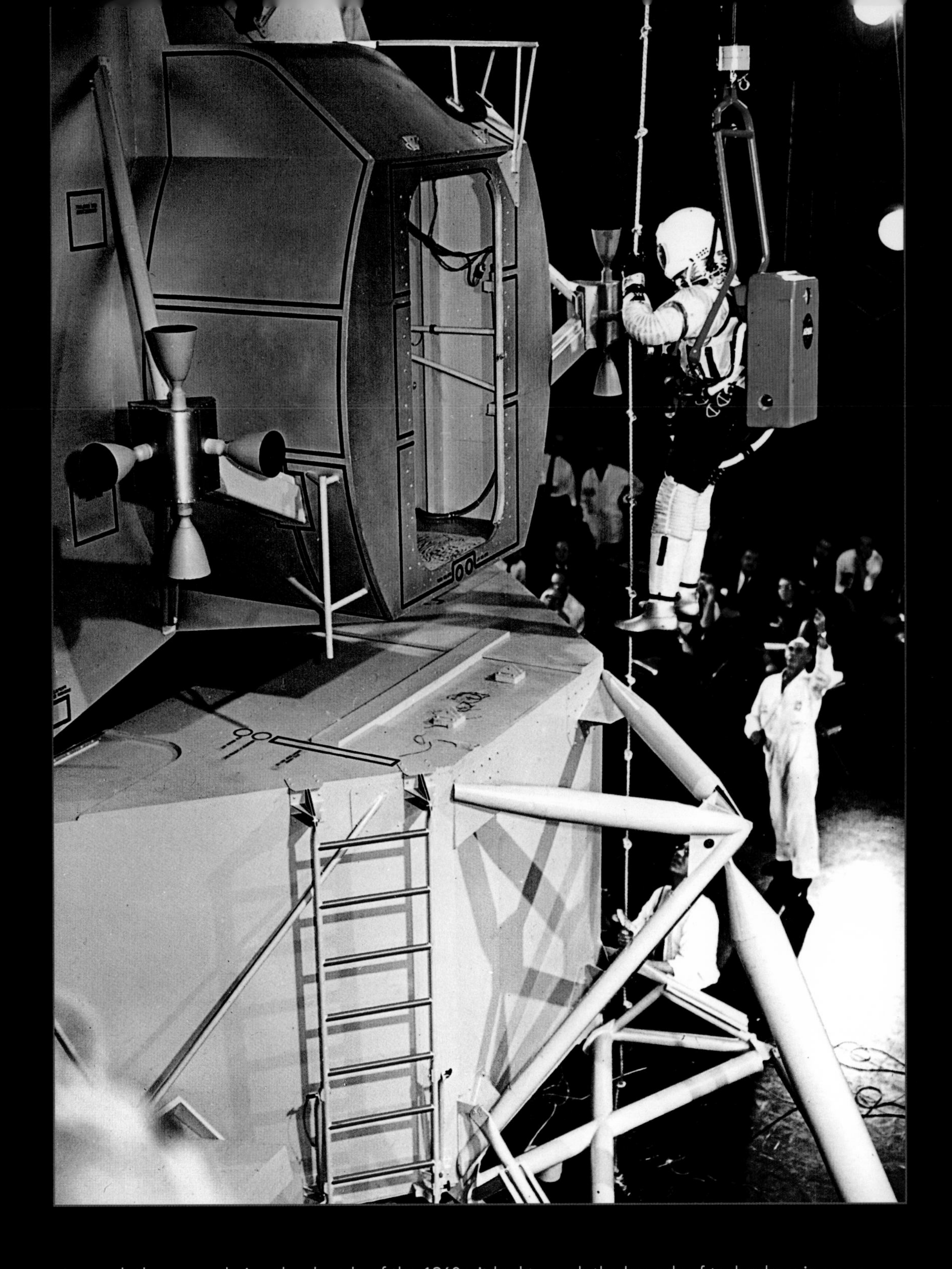

49

As NASA strove to reach the moon during the decade of the 1960s, it had to push the bounds of technology in many areas. The Lunar Module was the first true manned spacecraft. It would only operate in space, so it needed none of the sleek streamlining required by a machine operating in the earth's atmosphere. It was essentially composed of two boxes: a descent stage, with four legs, with an ascent stage sitting on it. This photograph shows an early mock-up of it. It was taken at the facility of the Grumman Corporation, the prime contractor, in 1964. At this time it was proposed that astronauts would descend to the lunar surface, and ascend from it, by means of a knotted rope. This idea was soon dismissed, following tests by astronaut Ed White and others. Instead, a ledge was built outside the hatchway and a ladder and handrail added on one of the landing legs.

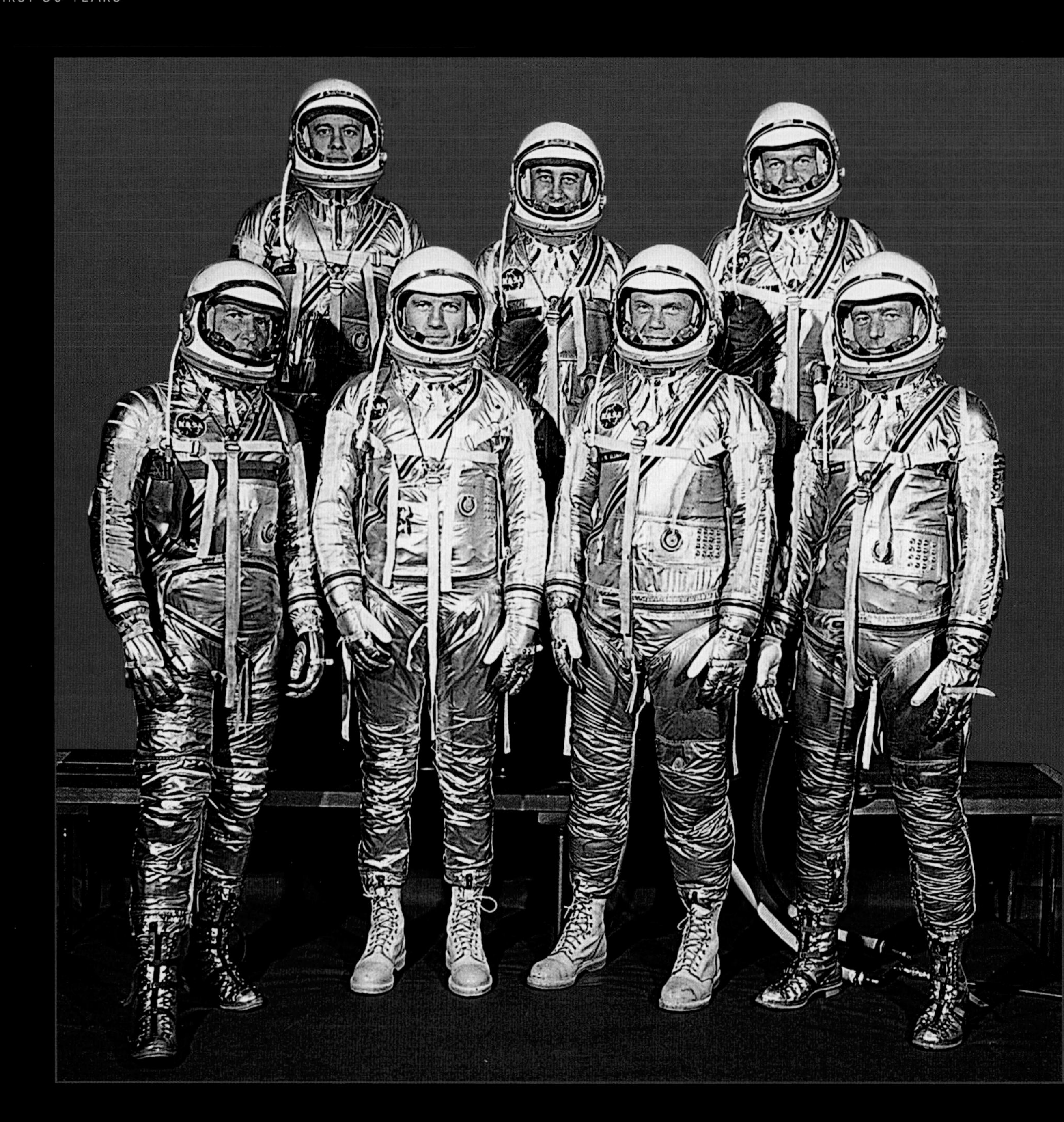

48

This picture recalls a time when space was the new frontier, in which the United States was trying hard to overtake the Soviet Union. NASA was formed in July 1958, and by April of the following year had announced the names of seven candidates for "astronauts," as space voyagers would be known. All were military test pilots, and all became public heroes, regarded as examples of American manhood at its best. They were not, of course, perfect, but only time would reveal that. They are shown here in their new and glamorous space suits, though these had to be greatly modified to become safe and serviceable. In the front row, from left to right, are Walter Schirra, Donald (Deke) Slayton, John Glenn, and M. Scott Carpenter. Behind them, left to right, are Alan Shepard, Virgil Grissom, and L. Gordon Cooper. All were to fly in space. Shepard landed on the moon in January 1971 as commander of the *Apollo 14* mission.

47

Look quickly at this image and it might be mistaken for the inside of a modernist chapel—with a stylized cross at the end. In fact, the two figures are not worshipers but engineers inspecting slosh baffles inside the liquid oxygen section of a space shuttle external fuel tank. During launch, the baffles reduce the movement of the fuel, helping to keep the vehicle balanced and under control. The overall length of the tank—comprising both liquid hydrogen and oxygen sections—is 154 ft. (47 m), and its diameter is 28 ft. (8.4 m). When this photograph was taken, the tanks were being built by Martin Marietta Aerospace at the NASA Michoud Facility in Louisiana. In recent years, the problem of debris falling from the protective foam cladding on the outside of the fuel tank has caused lengthy delays to the Shuttle program.

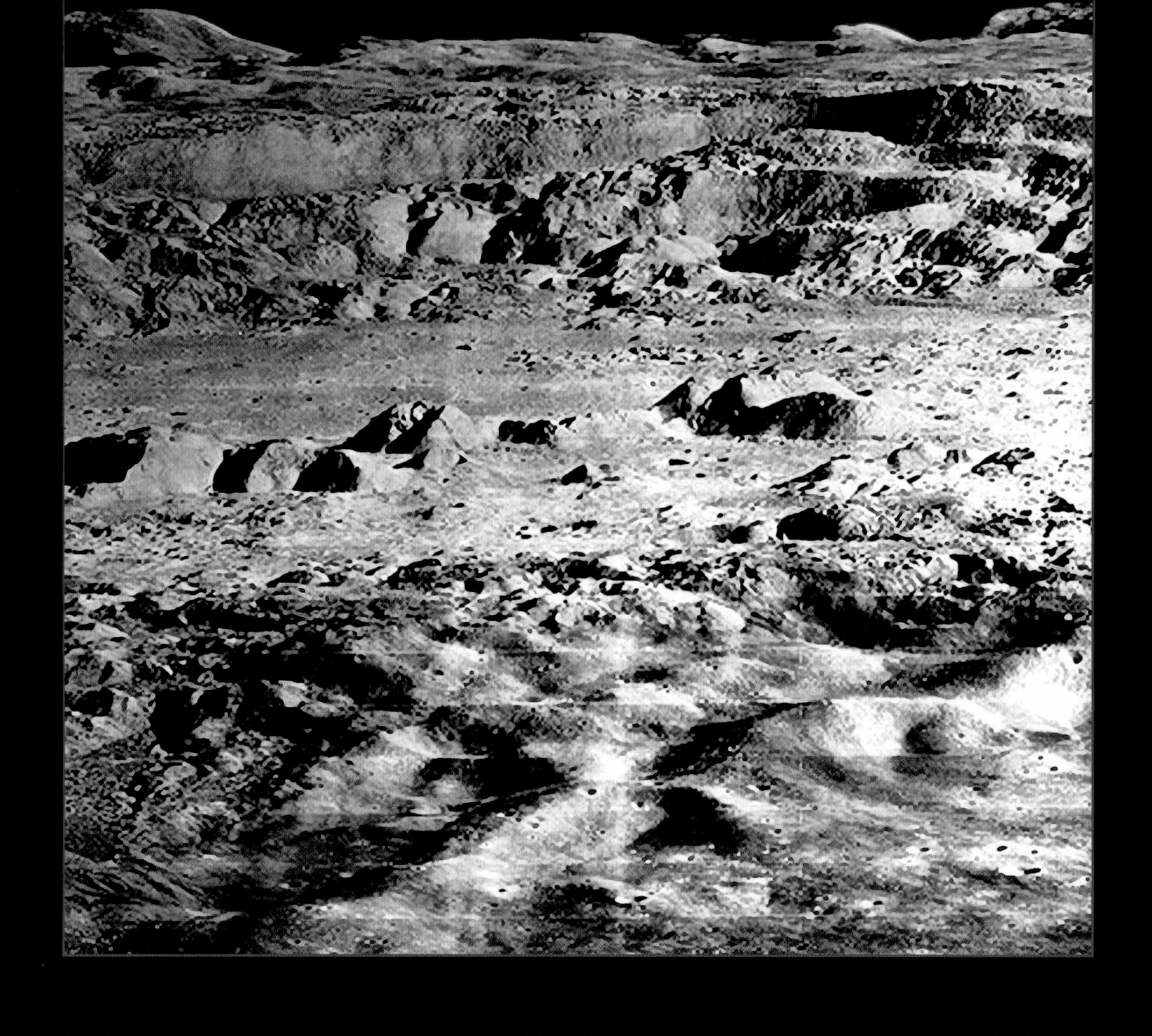

46

This image shows the lunar crater Copernicus, which is 60 miles (95 km) in diameter and about 2 miles (4 km) deep. When it was taken by *Lunar Orbiter II* in 1966 it was hailed as "the picture of the century." At the time, scientists and astronomers were hotly debating whether lunar craters resulted mainly from volcanic activity or from meteorite impacts. The argument has now been generally settled in favor of the bombardment theory. The central peaks of the crater would have been uplifted in a sort of rebound: thus the material forming them would probably have come from a great depth. The five successful *Lunar Orbiter* probes of 1966–7 were intended to map the lunar surface in preparation for the *Apollo* landings. They carried a camera loaded with film, which was processed on board. The film was scanned in strips and the information sent back to earth. This explains why most *Lunar Orbiter* images have lines running across them.

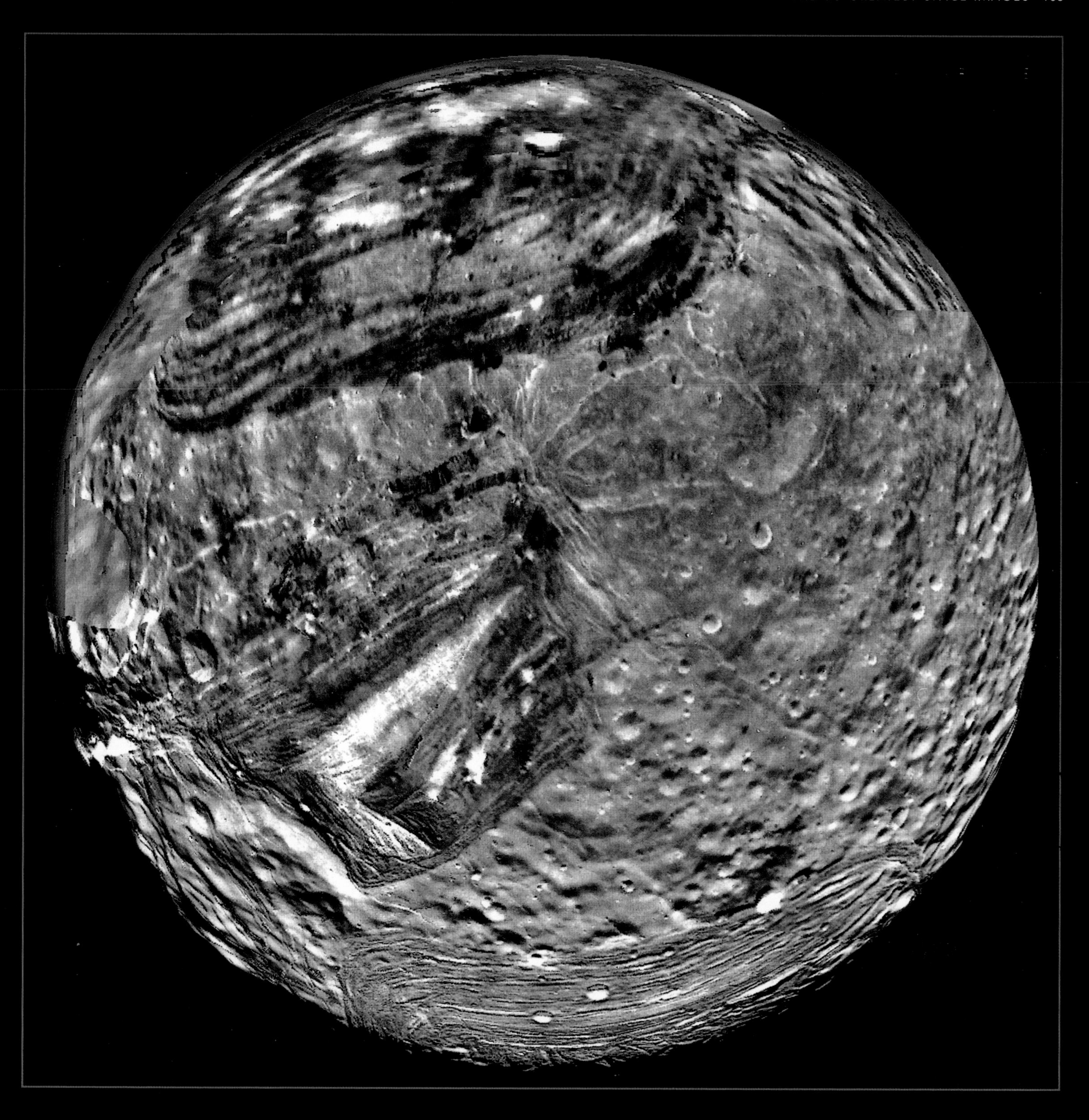

45

Miranda, a moon of the gas giant Uranus, is only 300 miles (500 km) in diameter but is one of the strangest bodies in the solar system. *Voyager 2* flew past Uranus in January 1986 and sent back many images of the planet and its moons. Early analysis showed the moons to be made of ice and rock. They are all named after literary characters, mainly from Shakespeare. This mosaic image of Miranda shows two distinct types of surface: old, heavily cratered, rolling terrain; and younger regions of terraces and what are believed to be fault canyons, some of them 12 miles (20 km) deep. These regions were later called "coronae." The most noticeable is the chevron at lower left of center. Two others can be seen, at the top and bottom of the picture. A possible explanation for them is that Miranda was once fractured by a violent impact, after which it was reformed under the influence of gravity.

44

The sun sets on the horizon as the crew aboard the space shuttle *Discovery* prepare to close the doors of the empty cargo hold at the conclusion of their mission in February 1997. Their main task had been to conduct five days of space walks, servicing the Hubble Space Telescope for the second time. A film camera was used to record this scene, which is heavily shifted to the red. Earth orbit missions have produced many dramatic horizon images, since the sun is frequently rising and setting for a spacecraft in a 90-minute orbit. At one extremity of this view is the typical twilight layering effect of red, white, and blue, which is caused by the refraction of sunlight passing through an extended distance of the earth's atmosphere

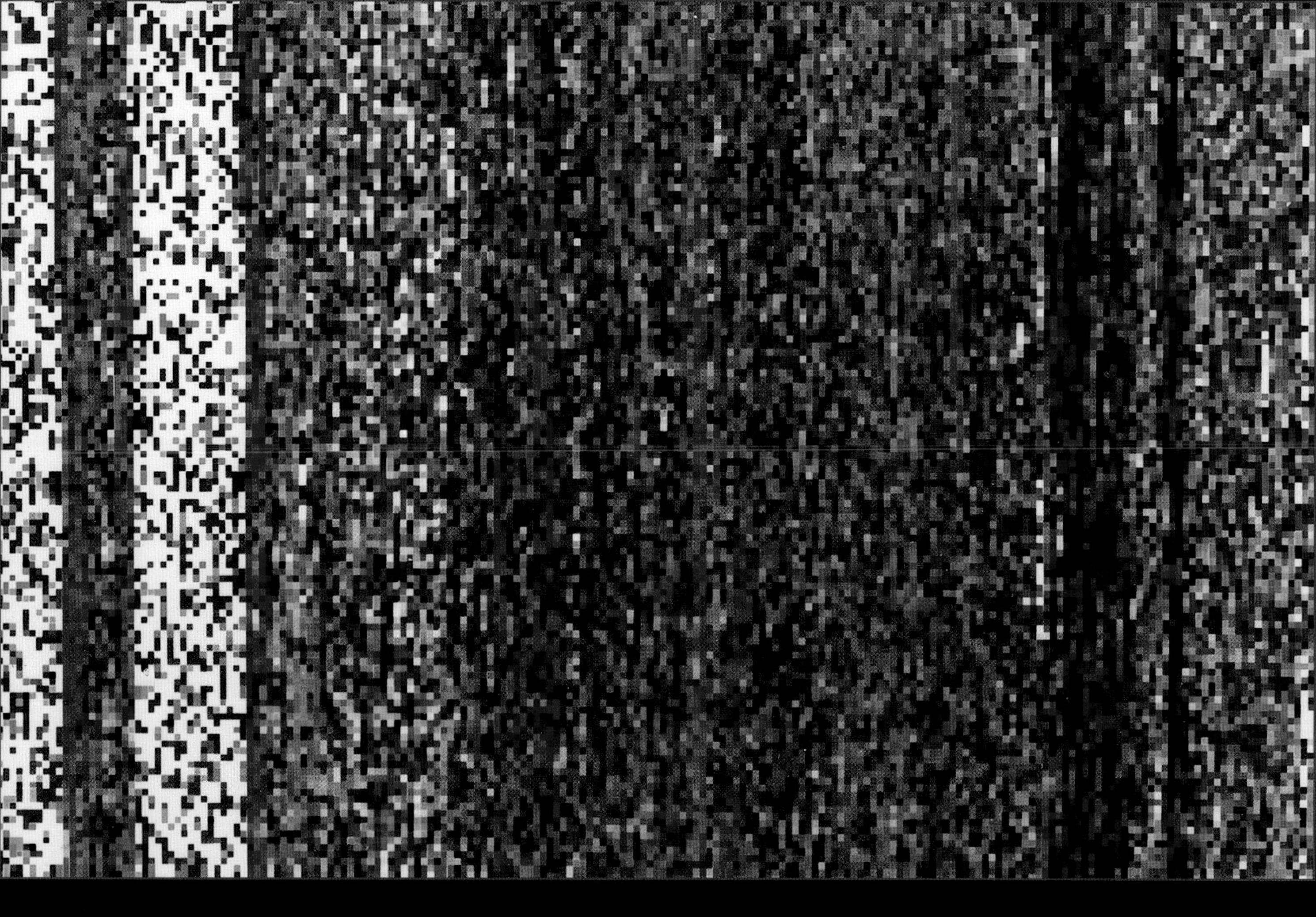

43

This image might be mistaken for the work of a modern abstract painter—perhaps a late work of Piet Mondrian. In fact, it shows color-coded data recorded between Baja California, Mexico, and Texas using a 10-channel multi-spectral infrared radiometer flown aboard one of the early Shuttle flights. The experiment was intended to test the value of infrared sensing from orbit in geological mapping. Yellow and green areas generally represent water. The first brown segment on the left is Baja California, and the second begins at the coast of mainland Mexico and extends into Texas. Dark brown strips on the extreme right are clouds. The colors of the individual picture elements, each representing 300 ft. (100 m) on the surface, indicate the type of ground cover—vegetation, rock, mineral, or soil. The distance on the ground represented by this image is almost 1,200 miles (2,000 km).

42

For the last three *Apollo* missions, a color television camera was fitted to the lunar rover so that viewers on earth could follow much of what happened during the extended geological traverses. This camera was operated from a console in Mission Control in Houston; it could be moved so that astronauts could be kept in view. At the end of a mission, the camera was pointed at the Lunar Module, and the lift-off of the ascent stage was seen as it separated from the descent stage (which acted as a launch platform) and climbed to an eventual rendezvous with the Command Module, which had remained in lunar orbit. Shown here is the first second or so of the liftoff of *Apollo 15* in August 1971. The bright patches of color spread across the image are reflections from pieces of the bright thermal cladding used to control temperatures within the Lunar Module as it sat on the lunar surface.

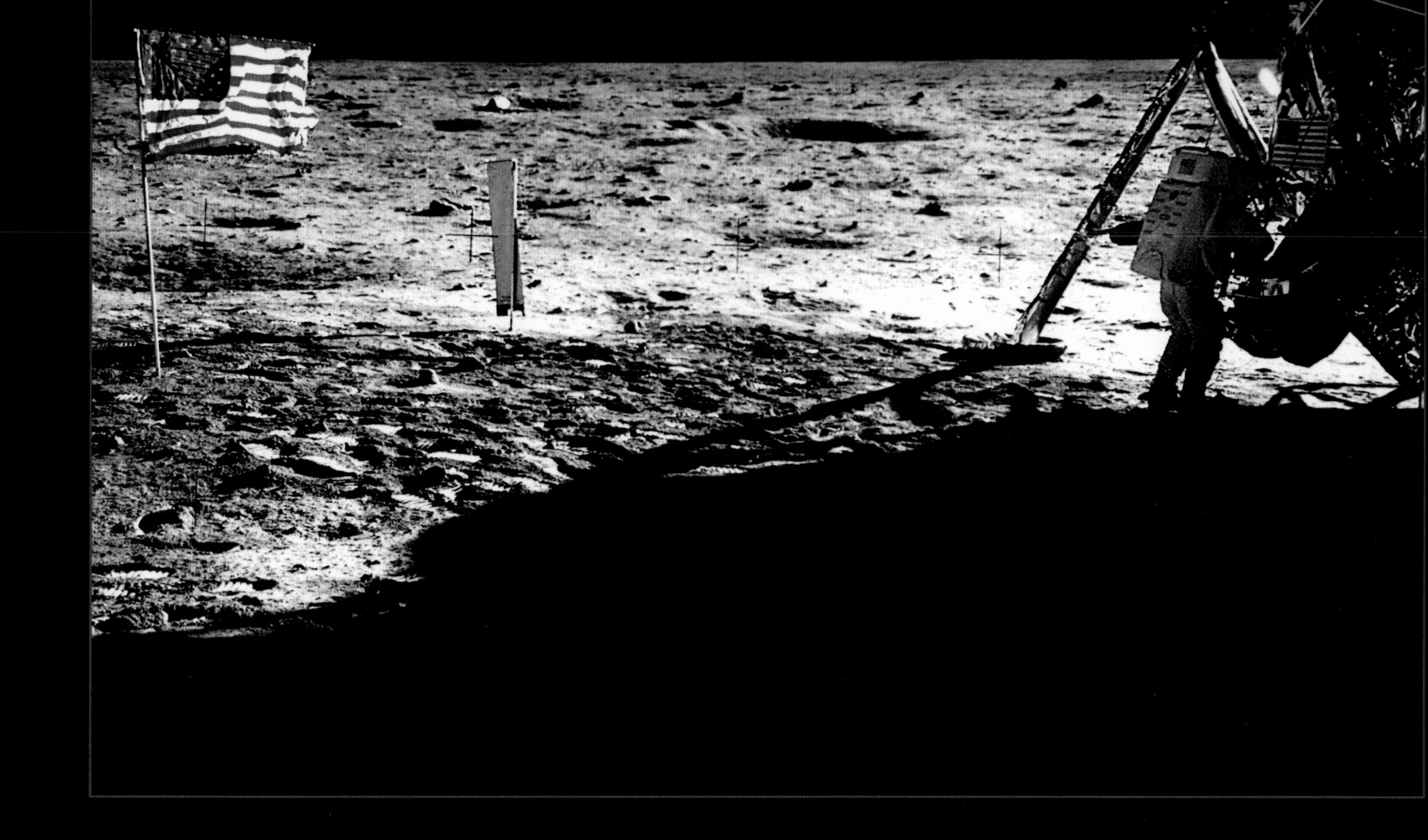

41

Mission Commander Neil Armstrong was the first man to step onto the moon, during the *Apollo 11* mission of July 1969. Surprisingly, this rather mediocre image is the only one that was taken of Armstrong on the moon. NASA—usually such a publicity-conscious organization—had somehow forgotten to instruct Edwin "Buzz" Aldrin, the Lunar Module pilot who followed Armstrong down onto the surface, to photograph him. This image is part of a panoramic sequence taken by Aldrin which happened to include Armstrong as he worked at the Lunar Module. Armstrong, on the other hand, had been instructed to take many photographs of Aldrin as he performed various tasks—and many of them turned out to be of excellent quality.

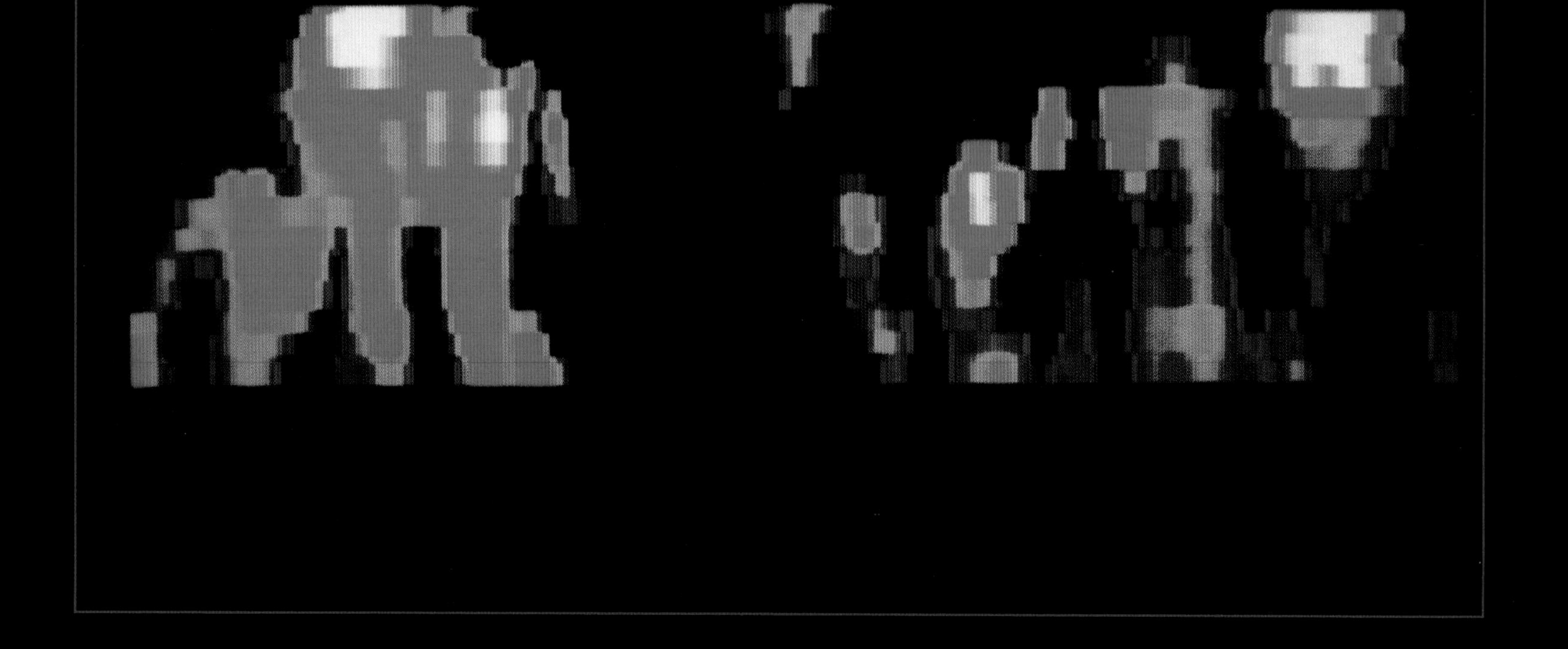

In 1983 the Infrared Astronomy Satellite mission was a shining example of international scientific cooperation—between the United Kingdom, the United States, and the Netherlands—and laid the foundation for extensive work in space-borne study of a region of the spectrum of great importance to astronomers. This image shows part of a bright nebula of gas and dust called 30 Doradus (or the Tarantula), which is 155,000 light-years from Earth in the Large Magellanic Cloud, the nearest galaxy to our own Milky Way. It is a portion of a scan across the Large Magellanic Cloud in which dozens of infrared sources, stars, and regions of dust and gas are seen, most of them not visible from the earth's surface. Some of the sources may be new stars forming behind clouds of dust and gas that radiate the stars' energy in the infrared.

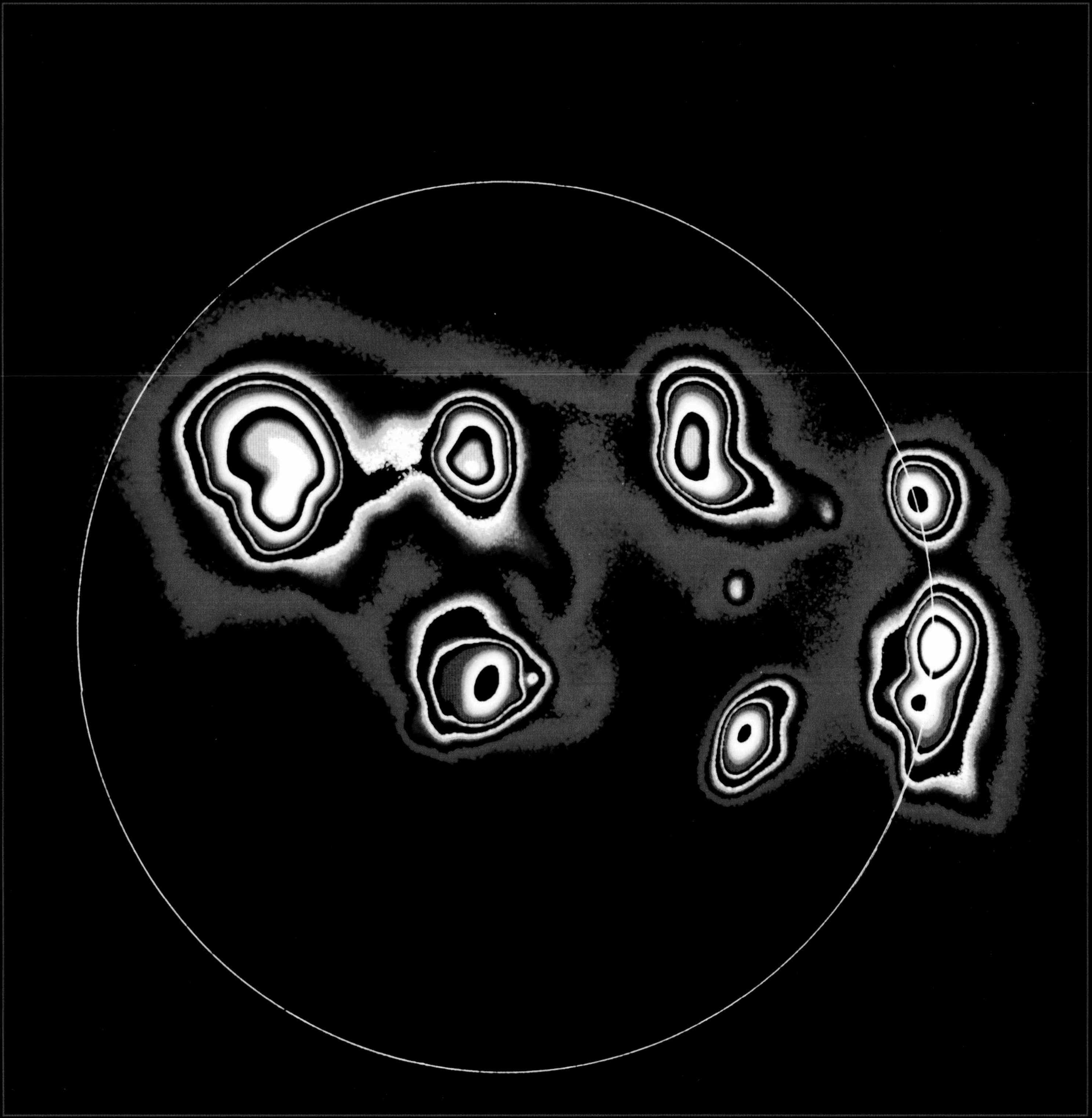

39

In 1973–4 the Apollo Telescope Mount, attached to the *Skylab* space station, yielded some 200,000 images of the sun in various wavebands. As well as advancing our knowledge of the processes at work inside our local star, many of these images were beautiful, mysterious, or exotic. This image was taken during a period when active regions were rotating around the solar disc. The picture, which records X-ray data, is a color-coded guide to levels of intensity in the active regions over a period of time. White corresponds to local temperatures of 9 million°F (5 million°C): a thousand times hotter than in the photosphere below, where the active regions appear as cool sunspots.

38

Over the course of years, missions into Earth orbit can track dramatic changes occurring on the surface below. A favorite of NASA crews has been the mouth of the Betsiboka River in the northwest of Madagascar. Clearance of brush and trees by farmers has exposed the soil to monsoon rains, with the run-off of red topsoil choking the river mouth. From space, reddish-brown water can be seen meeting the blue of the sea. Islands and mudflats have developed where there was once clear channel. This image, though, tells a different story. Taken from a high oblique almost facing the sun, it shows the coast around the Betsiboka etched by sun glint over the sea. The red here is not created by the soil run-off but by the sunlight. A close examination of several of the inlets does show the buildup of mudflats, however.

37

The advent of satellite technology has led to great improvements in the accuracy of weather forecasts. Weather satellites in both geostationary and polar orbits send enormous amounts of data back to earth—mostly in the form of numbers and black and white images, which may subsequently be rendered in color. Although weather systems are not generally the focus of manned orbital missions, they are a frequent subject for crews' cameras, and the results can be a valuable addition to the satellite record. For example, the eye of a hurricane can be recorded at different angles as the spacecraft passes overhead. This picture was taken over New Guinea by the crew of the STS-35 Shuttle mission in December 1990. The sun is setting and illuminating high clouds with a golden light. The scene may appear benign, but it includes a considerable number of storm clouds, which would be causing extreme conditions on the surface below.

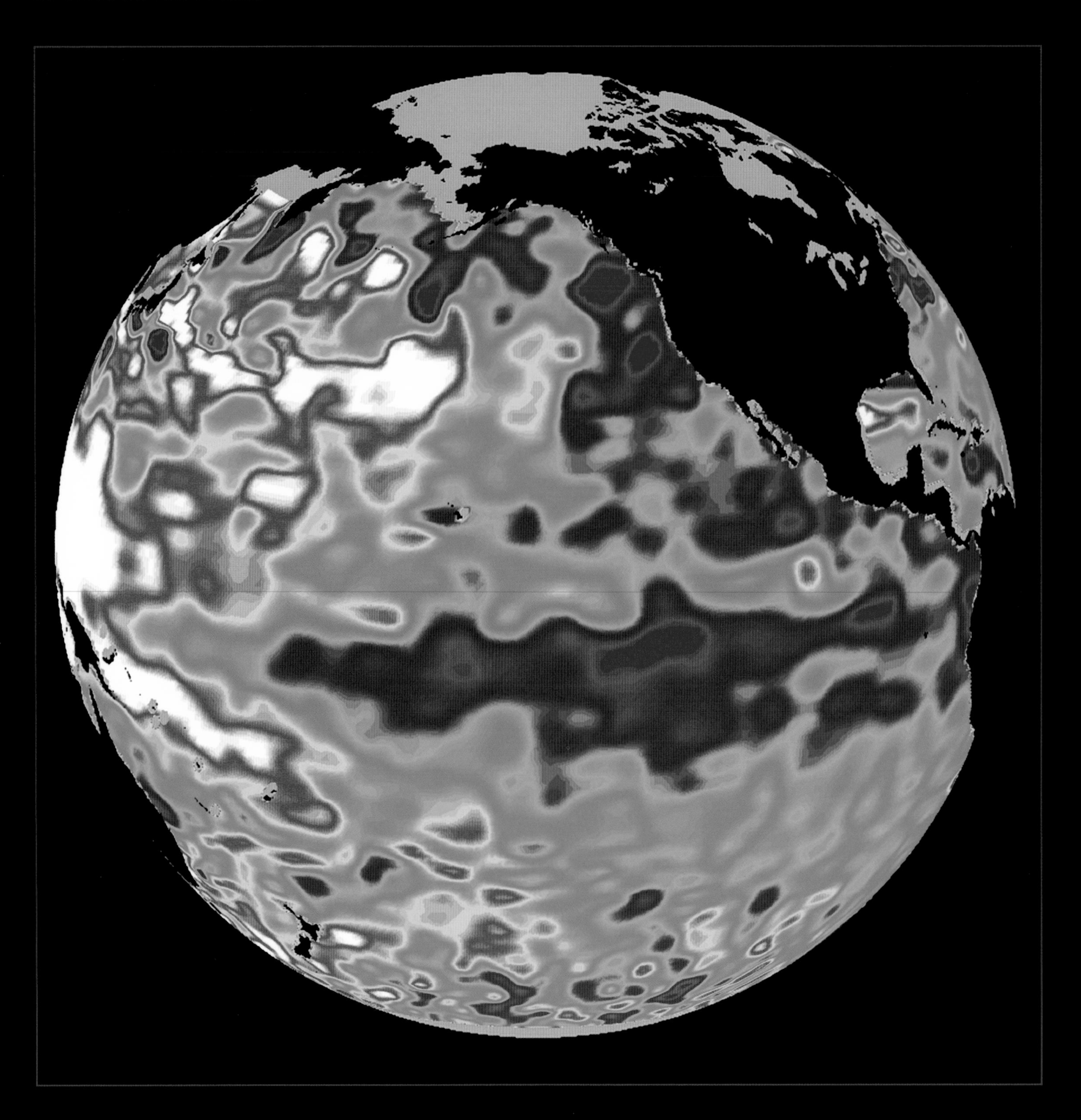

36

Sea temperatures in the Pacific Ocean have a significant effect on the world's climate. Two well-known phenomena—El Niño and La Niña—are marked by unusual water temperatures (warm and cold, respectively) in the Equatorial Pacific. In 1992 a joint U.S.–French satellite, Topex/Poseidon, was launched with a suite of instruments that measured sea surface heights every 10 days. Scientists were able to relate changes in sea height—which reflect patterns of heat storage in the ocean—to atmospheric climate. This false-color rendering of the Pacific was produced in March 2000, during a La Niña. Normal ocean height appears as green, with cooler water, at a height lower than normal, as blue and purple. A large area of warmer water in the western Pacific is shown as red and white. On the basis of such images it was predicted that the southern United States would have an unusually dry spring.

35

The crew of an alien spaceship approaching the night side of earth might conclude that the planet either had a natural energy source emitting light in many areas or an intelligent life form, which extended day into night by generating light. Since its inception in the 1960s, the U.S. Air Force's Defense Meteorological Satellite Program has been recording high-resolution night-time images. This view of Europe and North Africa is one example. While urban areas account for many of the locations colored a light blue, some of the light in North Africa and Arabia is flare-off in oil fields. The data gathered has been used to show that the growth of "mega-cities" is altering the landscape and atmosphere: by overlaying "light maps," which reveal urban sprawl, onto other data such as soil and vegetation maps, researchers have shown that urbanization can have a measurable effect on photosynthetic productivity.

34

Gemini XI—a three-day mission that flew in September 1966—set an altitude record of almost 740 miles (1,200 km) when it fired the engine of the Agena target vehicle to which it was docked. The crew, "Pete" Conrad and Dick Gordon, secured some memorable images of Earth. The image shown here marvelously illustrates the geological theory of "plate tectonics." It shows the area of the southern Red Sea and the Gulf of Aden. It shows clearly that—allowing for coastal changes over some millions of years—the Arabian landmass, if slid southwards, would fit well with the coast of Somalia below. In fact, the Afar Triangle immediately below the spacecraft forms a triple junction where three tectonic plates are pulling away from one another: the Arabian plate to the north and east, the African (Nubian) plate to the west, and the African (Somalian) plate to the south and east. The division between the latter two forms the East African Rift Zone.

33

In 2005 the *Deep Impact* spacecraft, after a journey of over 270 million miles (430 million kilometers), sent an impactor on a collision course with Comet Tempel 1. On July 4 the impactor smashed into the comet at a speed of 23,000 mph (37,000 km/h), before being vaporized. This image was taken 67 seconds after impact by a camera on board *Deep Impact*. Following the impact, which produced temperatures of several thousand degrees Celsius, a crater was formed on the comet's surface from which ice and dust were ejected. The image shows sunlight being reflected off the ejected material. The principal purpose of the mission was to gain information about the make-up of comets, which many believe to have been relatively unaltered since the formation of the solar system. NASA's *Stardust* spacecraft—after collecting particles of dust surrounding Comet Wild-2—yielded further evidence when it returned its cargo to Earth in January 2006.

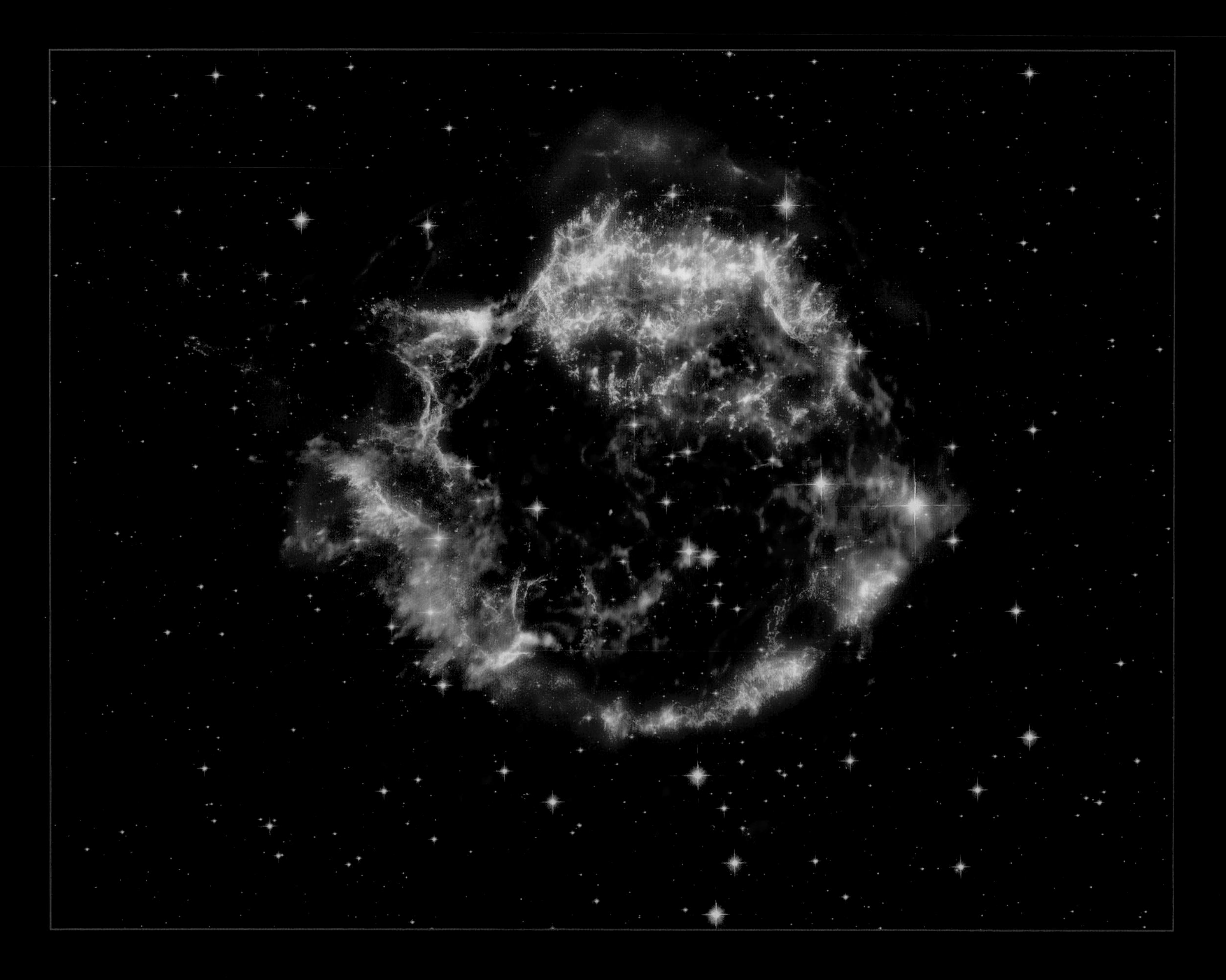

32

NASA has launched four "Great Observatories" into orbit. This image of the remnant of a 300-year-old supernova explosion, Cassiopeia A (Cas A), combines data from three of them. Optical data from the Hubble Space Telescope is colored yellow; infrared data from the Spitzer Space Telescope is red; and data from the Chandra X-Ray Observatory is green and blue. The Spitzer image reveals warm dust in the outer shell, with temperatures of about 50°F (10°C), while the Hubble image shows delicate filaments of hotter gases, around 18,000°F (10,000°C). The Chandra image records much hotter gases—about 18 million°F (10 million°C). These arose when material ejected from the supernova smashed into surrounding clouds of gas and dust at speeds of around 10 million mph (16 million km/h). Data from the telescopes may help to resolve the mystery of how cool dust grains can coexist with these super-hot gases.

31

The Hubble Space Telescope was launched into orbit from the space shuttle *Discovery* in April 1990. This impressive wide-angle image was taken just after the launch. The camera was located in the cargo hold and was operated remotely by the astronauts from their cabin. The Hubble story began inauspiciously when the first images returned to earth were found to be out of focus—the result of a manufacturing fault in the telescope's mirror that had gone undetected. Fortunately, optical correction was possible, and a unit to effect this was installed during the telescope's first servicing mission. The telescope went on to have a major impact on astronomical knowledge. There have been four Shuttle servicing missions so far, but it is still not clear whether a final mission to extend Hubble's life will be possible before the Shuttle is withdrawn from service in 2010.

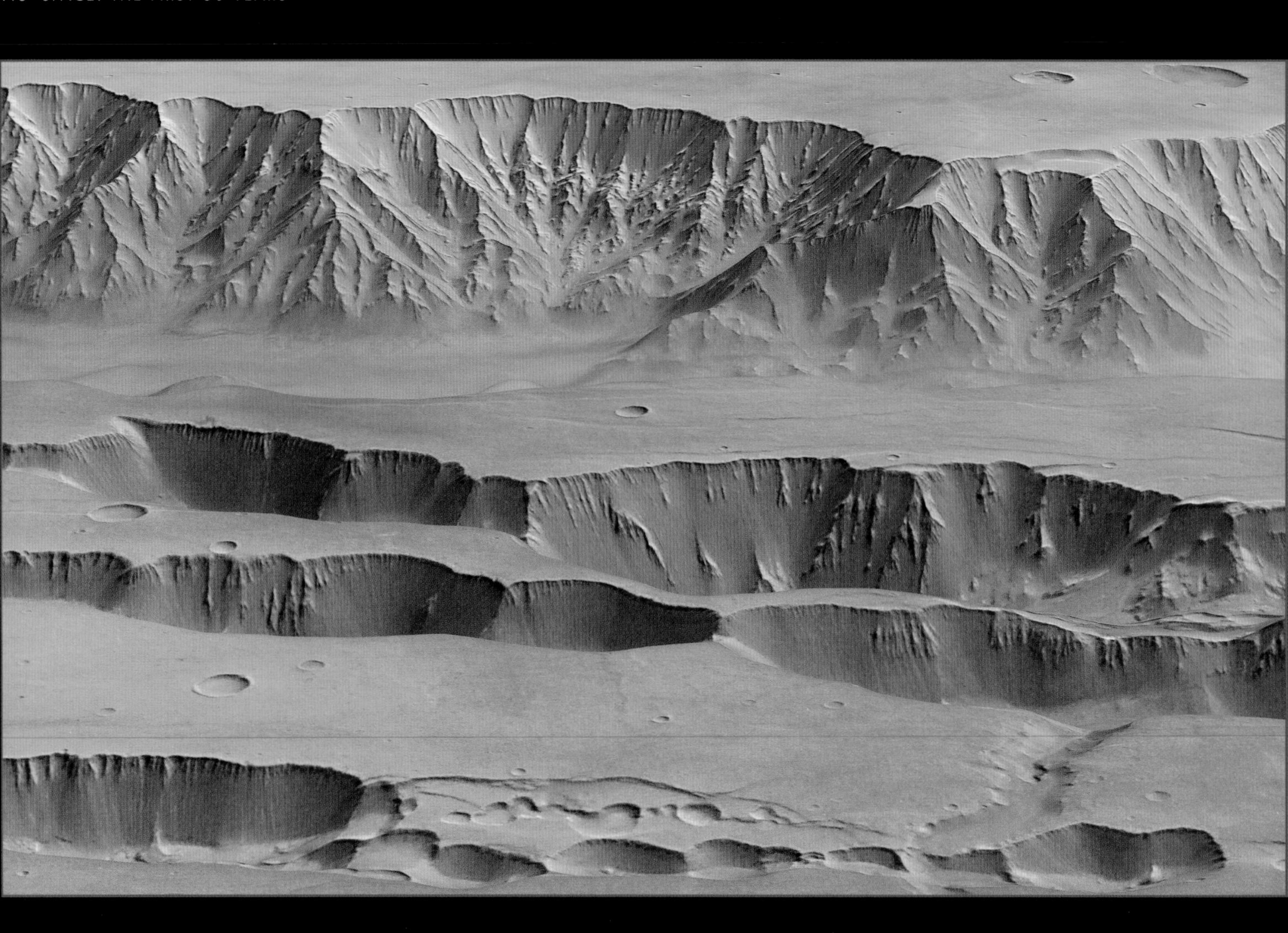

30

As the fiftieth year of the space era approached, attention focused on Mars. NASA's *Spirit* and *Opportunity* rovers were still performing beyond expectations on the surface, while three orbiters were active above—with a fourth scheduled to join them. One of the orbiters was *Mars Express*, ESA's first major interplanetary probe. Launched aboard a Russian rocket, it entered Mars orbit on Christmas Day 2004. When its nominal mission lifetime was completed, ESA decided to extend its activities for a full Martian year (687 Earth days) beginning in December 2005. The images released by ESA have been magnificent. This one shows a central region of the Valles Marineris canyon system. The main trough appearing in the top (northern) half of the image, Coprates Chasma, lies about 5 miles (8 km) below the surrounding plains. To the south of this is Coprates Catena, which comprises three less dramatic troughs, running roughly parallel, and up to 3 miles (5 km) deep.

29

Some images from space need more explanation than others. This Cassini image sent back to Earth from Saturn shows the shadows of the rings on the azure backdrop of Saturn's northern latitudes. Three images were taken—through infrared, green and ultraviolet filters—using the spacecraft's narrow-angle camera, from a distance of around 870,000 miles (1.4 million kilometers). These were combined to create this simulated "natural color" view. The planet appears blue because at the time the northern atmosphere was relatively cloud-free, with rays of sunlight taking a long path through it and thus being scattered at shorter (blue) wavelengths. At the bottom is Mimas, one of Saturn's icy moons, which is partially lit by the sun, revealing a cratered surface. The satellite is just under 250 miles (400 km) in diameter. Hidden from view is its huge crater, Herschel, which has a prominent central peak reminiscent of so many craters on the earth's moon.

28

Cassini-Huygens is a mission run jointly by NASA, ESA, and the Italian Space Agency, to study Saturn and its moons. Launched in 1997 on a lengthy journey past a number of planets from which it received gravity assists, it arrived at Saturn in July 2004. It has returned many magnificent images of the Saturnian system. When this image was released early in 2005, it was described, with some justification, as the greatest portrait of Saturn yet. It is a mosaic composed of 126 individual images obtained over the course of two hours from a distance of 3.9 million miles (6.3 million kilometers). The combined image is basically a "natural color" record. The sun is shining from the right, so part of the disc on the left is in shadow. Plainly visible are ring shadows cast on the blue-colored northern hemisphere. Later in the mission it is planned to obtain a similar image in which the planet is more fully illuminated.

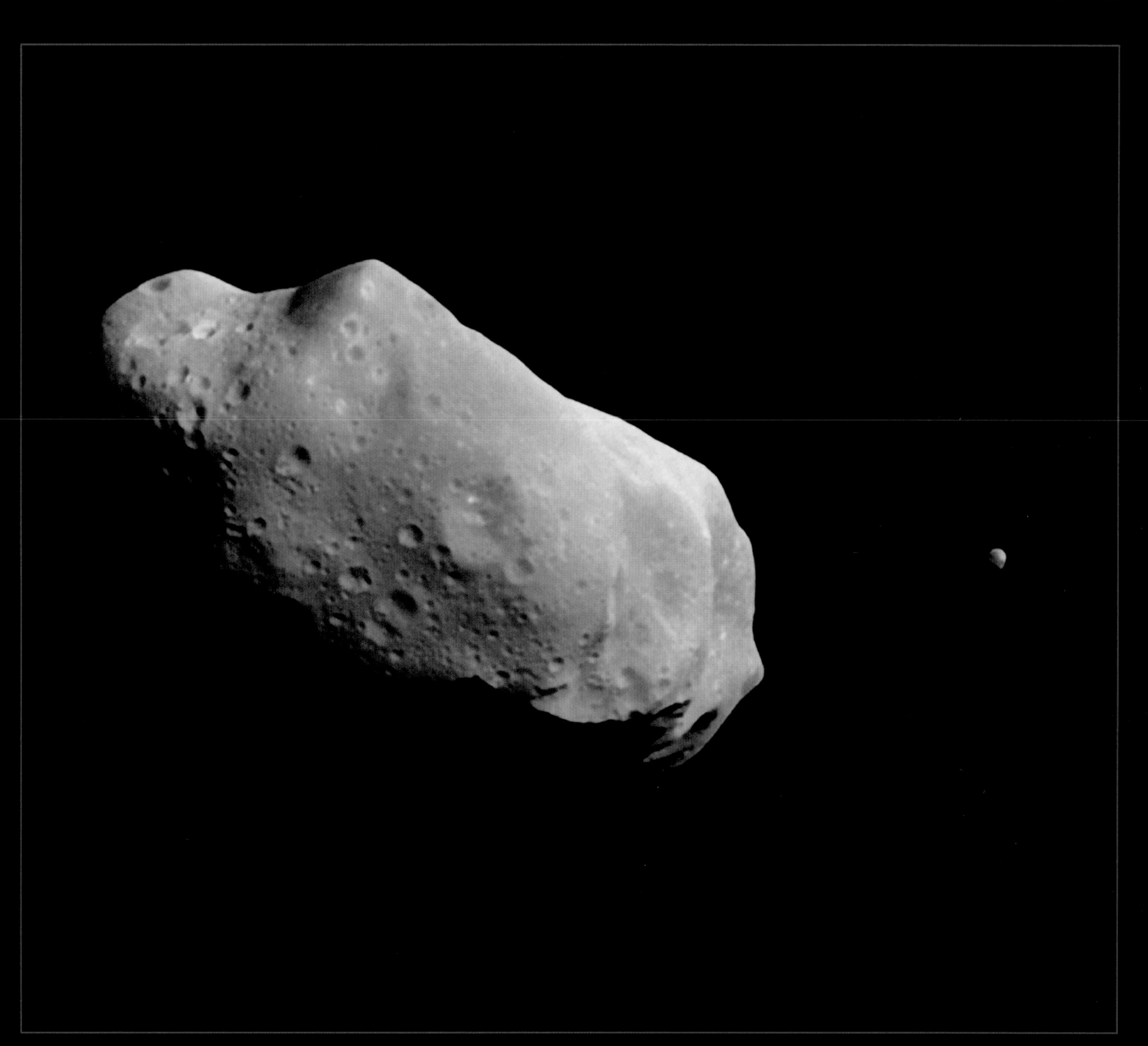

27

In October 1989 NASA launched a combined orbiter and descent probe to continue the detailed study of the solar system's largest planet, Jupiter. *Galileo* used gravity assists for its lengthy journey to Jupiter—from one fly-by of Venus and two of earth—and made numerous discoveries, including strong evidence that the moons Europa, Ganymede, and Callisto all had layers of liquid salt water. On the way to Jupiter, *Galileo* flew close to two asteroids, Gaspra and Ida—the first such visit by any spacecraft. This is a color composite of images of Ida taken by *Galileo* on August 28, 1993, from a distance of 6,500 miles (10,500 km). The images revealed a moon orbiting Ida (seen at the right), subsequently named Dactyl. The asteroid, which shows obvious signs of heavy cratering, exhibits many subtle differences in its rock: there is still active research into how much asteroids may have changed since their creation early in the history of the solar system.

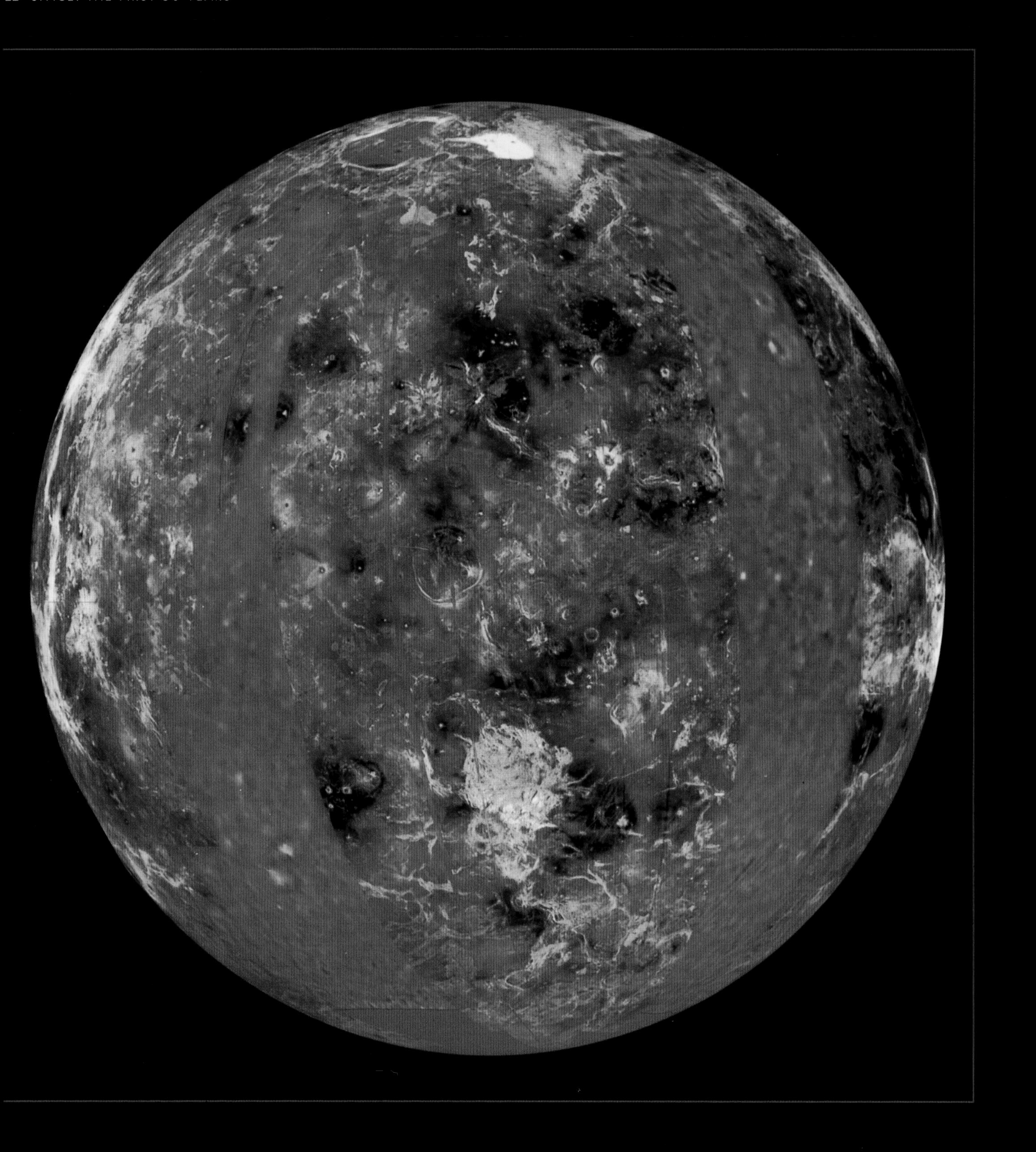

26

In folklore Venus has been regarded as earth's twin. In fact, the first probes to the planet revealed a hellish climate, with crushing pressures, searing temperatures, and clouds composed of sulphuric acid. It needed spacecraft equipped with radar imaging to see through the clouds, and it wasn't until NASA's two Venus *Pioneer* probes of the late 1970s that a global picture began to emerge. After much delay, the sophisticated Venus *Magellan* spacecraft was eventually launched from the space shuttle *Atlantis* in May 1989. This image is one of the early full-disc radar images produced by *Magellan*. In it, brighter areas correspond to higher elevations and the simulated hues are based on data from Soviet *Venera* landers. The bright area at the top is the Ishtar Terra region. *Magellan* orbited Venus for four years, mapping over 99 percent of the surface, before it plunged to destruction in [illegible] October 11, 1994.

25

When Halley's comet visited the inner solar system in 1985–6, it was studied in great detail. The most successful Halley mission was the European Space Agency's *Giotto* probe, which flew past the comet in March 1986. This image, a composite of seven frames, was the most impressive image of a comet ever obtained. A color camera recorded craters and elevations on the surface of the 10 x 5 x 5 mile (16 x 8 x 8 km) potato-shaped nucleus, which was one of the darkest objects encountered in the solar system, reflecting no more than 4 percent of the sunlight falling on it. Bright dust jets or filaments are a prominent feature of the image, in which the sun is shining from the left. Although subsequent study has suggested some modification of our ideas about comets, the results from *Giotto* and other Halley missions basically confirmed the "dirty snowball" model of the nucleus, first proposed in 1950 by the American astronomer Fred Whipple.

24

At NASA's Jet Propulsion Laboratory during encounters of the *Voyager* spacecraft with the gas giants of the outer solar system, tremendous excitement greeted the arrival of every new image, which then had to be interpreted at a few minutes' notice for the assembled crowd of journalists. This image was taken on February 5, 1979, when *Voyager 1* was 17.5 million miles (28.4 million kilometers) from Jupiter. The moon Io—which was shown to be volcanically active—is seen against the disc of Jupiter, while to the right is the icy moon Europa. At the bottom of the frame is Callisto, over four times further away from Jupiter than Io and much larger than Earth's moon. These three moons, together with the largest, Ganymede, are known as the Galileans because they were seen by *Galileo* in 1610 using his newly invented telescope. The most prominent feature of Jupiter's banded atmosphere is the Great Red Spot, a major atmospheric disturbance.

23

Mars has fascinated people for thousands of years. It is a challenging target for space missions, with successes alternating with failures. But the Viking project, in the 1970s, demonstrated what could be achieved. Two landers were sent to the

22

This is one of the antennae of NASA's Deep Space Network operated by the Jet Propulsion Laboratory in California
Sending sophisticated spacecraft to the depths of the solar system is one thing, but such missions would be pointle

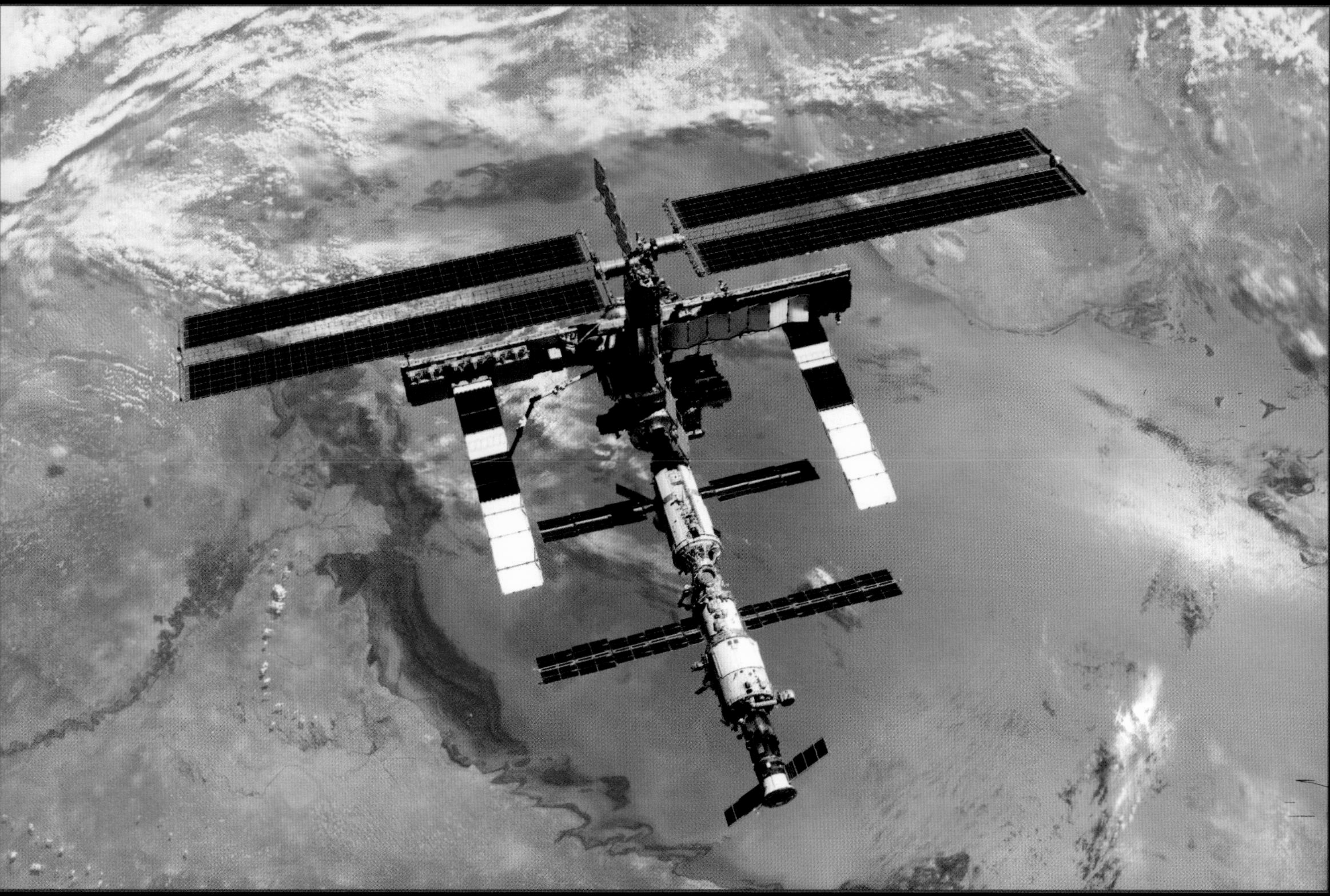

21

The *International Space Station* has had a checkered history. Envisaged as a major step forward in international space cooperation—between the United States, Russia, Europe, and Japan—its assembly and operation have been severely affected by the problems encountered by the space shuttle. It has been operated by a much-reduced crew of two, and the project has owed much to Russia's continuing ability to keep the station supplied with essentials by means of its automated space vehicles, and to ferry crew members aboard *Soyuz* spacecraft. The last shuttle flight to the *ISS*—before renewed concerns about potential damage to the orbiter from fuel tank debris during launch—was that of *Discovery* in August 2005. As their mission neared its end, the astronauts aboard the shuttle recorded images of the space station seen against a backdrop of the Caspian Sea. The large solar arrays—and a *Soyuz* spacecraft docked to the trailing end—are clearly visible.

20

The space shuttle was first conceived as a "space truck," capable of handling all kinds of cargo—particularly satellites. This bold concept has since had to be abandoned: the remaining orbiter missions are to be dedicated to completion of the *International Space Station*, before the shuttle is phased out altogether. But in 1984, when this image was taken, the shuttle was still regarded as a "jack of all trades." The crew aboard the *Discovery* orbiter launched two communications satellites; but their most difficult task was to retrieve two other satellites that had been launched earlier in the year but placed in incorrect orbits. Two spacewalking astronauts used a powered maneuvering unit and a specially designed "stinger" to lock on to the errant satellites and bring them close enough to *Discovery* for the orbiter's remote arm to place them in the cargo hold for return to Earth and refurbishment. Such missions demanded great skill and bravery of the astronauts.

19

The reentry of a manned space vehicle from orbit is a critical phase. The friction caused by the earth's atmosphere raises temperatures on the outside of the vehicle to at least 3,000°F (1,600°C). This heat can be dissipated by the use of an ablation shield on the front of the spacecraft: a coating intended to burn away, leaving the crew safe inside. This method suited spacecraft that were not going to be used again, but the space shuttle required a more sophisticated system of tiles and blankets fixed to the outer structural skin of the orbiter. During the early years of space flight, few images of reentry were taken from a crew compartment—perhaps because of the cramped conditions—but some good images have been obtained from shuttle missions. In this one, the large forward-looking windows of the flight deck show the hot plasma being generated around the vehicle as it encounters the atmosphere.

18

From the beginning of the space era, rocket launches have been one of its most dramatic facets. Nothing could rival the awe-inspiring nature of a Saturn V launch during the *Apollo* era, but shuttle launches since 1981 have come close. During *Apollo* we typically saw wide field images of the spectacle from the side together with close-ups of the flame trench under the launcher stack as well as the climb out of the vehicle past the top of the launch tower as liftoff took place. An additional and powerful angle was added when the space shuttle went into service. Cameras just to one side underneath the shuttle, which was mounted on its large fuel tank and solid rocket boosters, recorded its ascent and gave perhaps the best impression to date of "being there." This is a classic example of the genre—with *Columbia* heading for the sun lit cloud deck above at the beginning of the STS-50 mission in June 1992.

17

Sometimes fantasy foreshadows reality. This image might look like a still from Stanley Kubrick's classic film *2001: A Space Odyssey*. In fact, it is a spacewalking shuttle astronaut testing a new device for operating outside a spacecraft without the constraints of a safety tether. The "manned maneuvering unit" was worn on the back of the astronaut's space suit, and gaseous nitrogen propellant enabled the astronaut to move in three dimensions at a controlled speed. Once in the desired position, an "attitude hold" function could be engaged, allowing the astronaut to work without having to operate the hand movement controls. The device was used during three shuttle missions in the mid-1980s. This image shows Bruce McCandless at a distance of about 300 ft. (100 m) from *Challenger* during the first of those missions, STS-41B, in 1984. A smaller unit intended primarily as a safety device for use on the space station was flown on a later Shuttle mission in 1994.

16

Many scientists argue that automated probes can achieve anything an astronaut can, and more, at lower cost. But manned missions can be better suited to dealing with the unexpected. *Skylab*, NASA's first space station, was launched in May 1973. Soon afterwards it was found that a shield, designed to protect against meteoroids and control temperatures within the station, had been torn off, along with one of the two major solar arrays; the other array had been damaged. In due course a three-man mission travelled to the station, where it found that the second solar array could be unjammed during a space walk. An improvised canopy was deployed, bringing temperatures within the vehicle down. This is perhaps the best image showing *Skylab* in orbit, battered but unbowed. Between May 1973 and February 1974, 3 three-man crews spent 28, 59, and 84 days respectively aboard *Skylab*, performing important scientific experiments.

15

Early in an Apollo lunar mission, the Command Module had to withdraw the Lunar Module from the top of the third stage of the Saturn launcher, where it had been carefully protected during liftoff. *Apollo 7*, in October 1968, was not a lunar mission, nor was the Lunar Module flown, but this extraction maneuver was one of the procedures rehearsed. Here, the white disc within the open panels of the third stage replicates the target that was to appear on the top of the Lunar Module. Below the cruciform shape of the target vehicle the envelope of the earth's atmosphere can just be seen. Beyond the vehicle there appears to be a vista of stars. But in space, things are not always as they appear: in fact the "stars" are frozen particles of expended fuel and urine that have been dumped from the spacecraft—creating, as one astronaut is said to have quipped, the "constellation of Urion."

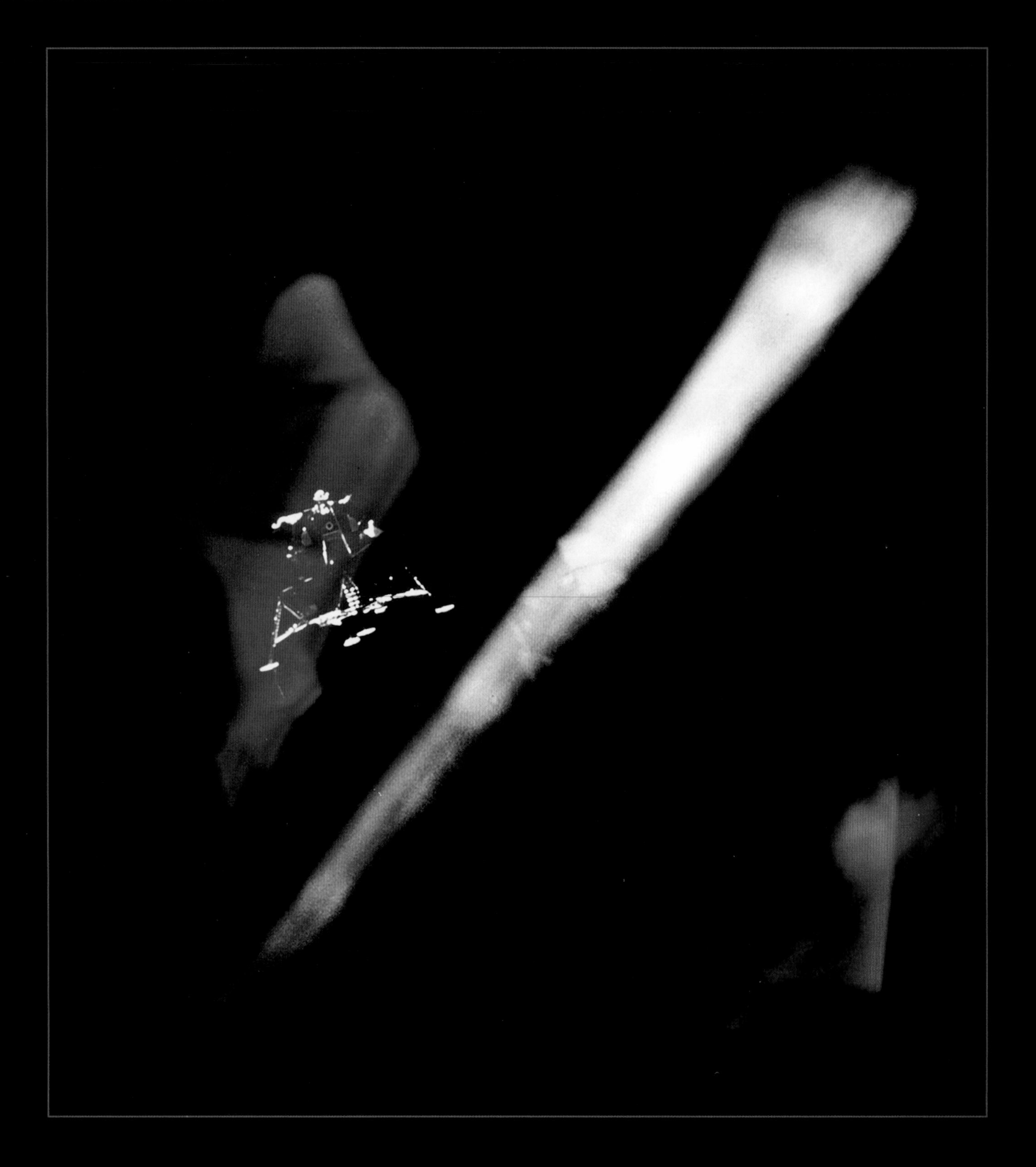

14

Some space images are direct observations of a particular subject; others are created compositions; and yet others might be said to result from chance. This image may be placed in the last category. In May 1969, after the successful test of the new Lunar Module in Earth orbit during the *Apollo 9* mission, NASA decided to test the spacecraft in lunar orbit—with virtually all of the procedures required for a lunar landing mission, apart from the actual landing, being carried out. The four-legged vehicle with astronauts Tom Stafford and Gene Cernan on board separated from the Command Module and was photographed by Command Module pilot John Young. In this frame, reflections from lights inside the Command Module on its windows create what look like exotically colored lunar nebulae around the stark but ghostly shape of the Lunar Module—an example of reality becoming fantasy.

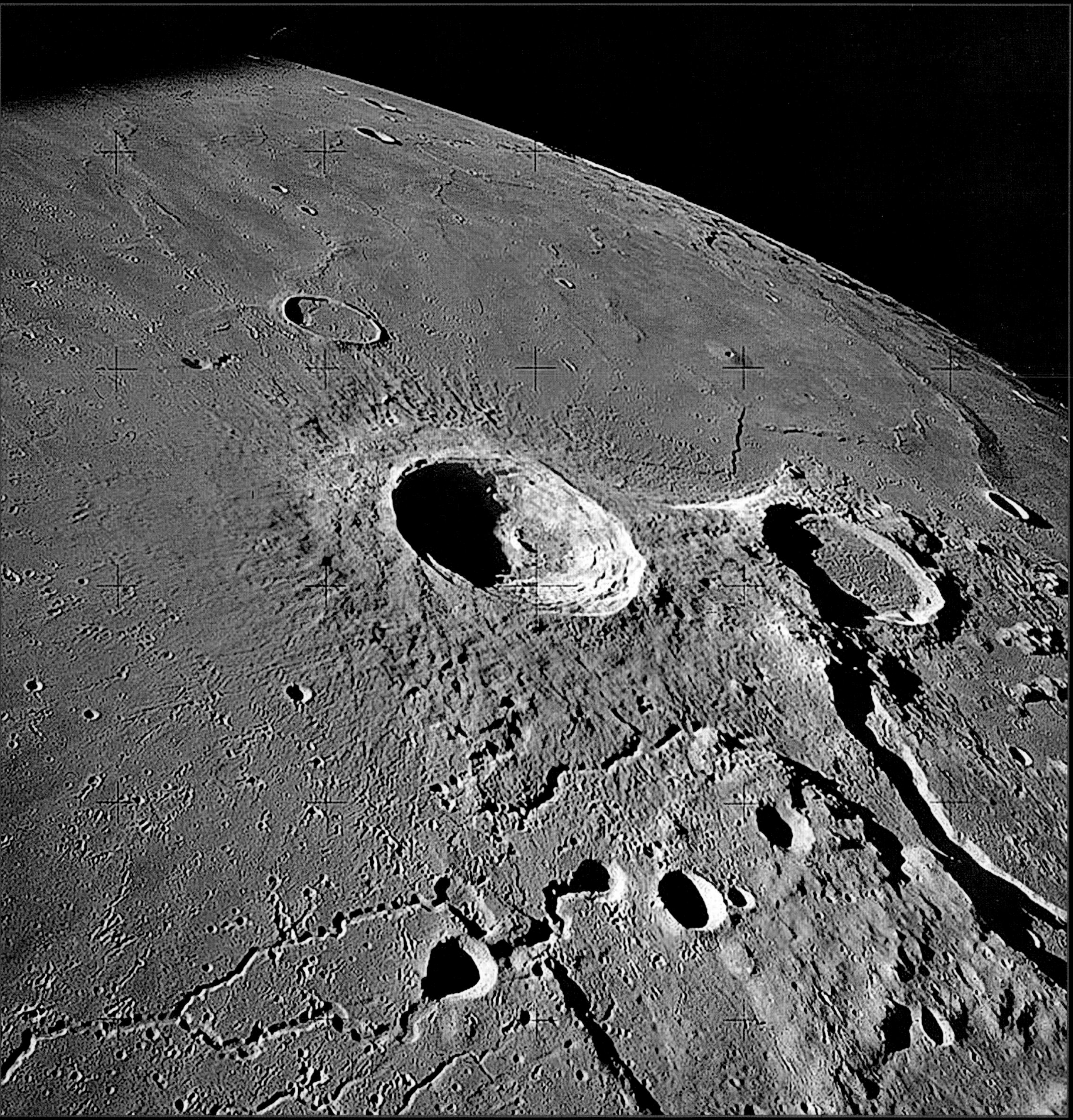

13

Astronomers on Earth have to view the moon through the earth's atmosphere, which has a distorting effect. It was a revelation when unmanned spacecraft and Apollo missions went to the moon and took images of its surface from close up. This *Apollo 15* image—a black and white rendering of an original color frame—demonstrates the detail that was recorded. It shows a particularly dramatic region of the moon. The large crater in the middle is Aristarchus, an extremely bright feature that is the center of a distinctive ray system and which can be seen from the earth in Earthshine, when lunar night has enveloped the area. The other major crater is Herodotus. The valley on the right and running out of the frame is Vallis Schröteri, the largest sinuous valley on the moon. Its bulbous head, immediately south of Herodotus and seen in heavy shadow here, is a volcanic crater known as the Cobra

12

Many *Apollo* pictures were quite basic: shot simply to document scientific experiments. Others were well-observed compositions taken during periods of intense activity. This image from *Apollo 17*—the last lunar mission—taken by mission commander Eugene Cernan shows Lunar Module pilot Harrison Schmitt working by the lunar rover vehicle close to the landing site in the Taurus-Littrow region of the moon. The picture sums up what one astronaut called the "magnificent desolation" of the lunar surface. The scene is totally unlike a typical view on the earth: there are no trees, telephone poles, or any other objects to give a sense of scale, other than Schmitt and the rover. The horizon could be one or ten kilometres away. The isolation of the human being in this alien landscape emphasizes his vulnerability and underlines the achievements of the Apollo program in not only exploring a new world but bringing back all crew members safely.

11

Many memorable images from the space era achieve their effect thanks to the intrinsic drama of an event. A much smaller number are moments captured spontaneously by the "seeing eye" of a perceptive astronaut. There are very few "portraits," but this picture of *Apollo 7* astronaut Walter Cunningham belongs in that category. The face is seen in a sharply etched silhouette, lit by direct sunlight at the front and in deep shadow beyond. Cunningham's head is framed by one of the Apollo Command Module's windows and by communication cables. His features might be taken as portraying the extremes of spaceflight—stubble on the chin suggesting the harsh physical demands, the expression suggesting a brief, quiet moment taken to look out of the spacecraft into the enigma of space.

10

The visual story of the space era is frequently punctuated by images of rocket launches—and particularly those of the awe-inspiring Saturn V launcher which, standing over 360 ft. (110 m) high and developing 7.5 million pounds (31 million kilograms) of thrust, sent Apollo crews on their way to the moon. Anybody who witnessed an Apollo launch would remember the experience forever: from 3 miles (4.8 km) away the ground shook violently, the body was buffeted by shock waves and the ripping, eardrum-tearing noise was indescribable. The Saturn V was dwarfed by the "stack," the mobile launcher that served both as an assembly platform during the rocket's construction and as a launch platform. Before a launch it was lit for a considerable period by bright searchlights. Even at rest, the Saturn was awe-inspiring: an example of the fabulous architecture of the space era, shown so impressively in this image.

09

Astronauts Tom Stafford and Walter Schirra aboard *Gemini VI-A*, and Frank Borman and James Lovell inside *Gemini VII*, conducted the world's first rendezvous in space on December 15, 1965. This image shows the *Gemini VII* station keeping close to the second spacecraft, which had been launched less than six hours before. The two spacecraft flew in tandem at various distances from each other for three orbits of the earth before they separated. *Gemini VII* returned to earth after little more than a day in orbit. Borman and Lovell, however, still had three days of their trail-blazing two-week mission to endure before they could return to earth and escape from the confines of the capsule.

08

Many of the procedures that were vital to the success of the Apollo project—such as the rendezvous and docking of spacecraft, and the space walk—were established during the two-man *Gemini* missions of 1965–6. This picture shows the first spacewalk (EVA, or "extravehicular activity," in the jargon) by American astronaut Ed White on *Gemini IV*—10 weeks after the world's first, by Soviet cosmonaut Alexei Leonov in June 1965. The astronaut floats beside the spacecraft and is tethered to it by an 25 ft. (8 m) "umbilical cord" which delivers oxygen from the spacecraft for breathing and suit pressurization. In his right hand White holds a small gas-powered maneuvering unit; mounted on it is a 35mm camera. White spent almost 21 minutes space walking. He found it exhilarating but exhausting. Most of the problems were resolved by procedures adopted for the final *Gemini* mission at the end of 1966.

07

Many impressive images of the earth from deep space were taken by *Apollo* crews traveling to the moon, but almost all featured an Earth in a crescent or gibbous daylit phase. This is the only virtually full-disc Earth obtained—by the crew of the final Apollo mission (*Apollo 17*) in December 1972. The African continent is plainly visible. Perhaps the most interesting feature is a demonstration of why the earth has seasons: with the planet facing the sun, at the time of the southern-hemisphere summer, a large area of Antarctica is visible and the northern hemisphere is tilted away. Major cloud (storm) systems can be seen just south of Africa, and there is much cloud occupying the region of the equator further north. Lunar Module pilot Harrison Schmitt, the only professional geologist to travel to the moon, commented at great length during the outward journey about the earth's meteorology, so vividly displayed to the space traveler.

brilliantly successful. Many people still vividly remember the crew reading from the Book of Genesis during one of their 10 two-hour orbits around the moon as Christmas neared. The crew took the first-ever color images of an Earthrise above the lunar horizon, and mission commander Frank Borman commented to mission control in Houston: "We were just saying that there's a beautiful Earth out there."

Space flight is a highly dangerous endeavor, and claims that orbital flight might one day become "routine" are nonsense. In January 1967 the United States's *Apollo* moon project suffered a major setback when the three-man crew of *Apollo 1* died in a Command Module fire during a test at the Kennedy Space Center. Several months later, Soviet cosmonaut Vladimir Komarov died as his spacecraft returned from orbit. Both of these tragedies occurred away from the public gaze. But that of the space shuttle *Challenger* on January 28, 1986, occurred with the world watching. A fault in one of the launcher's solid rocket boosters led to an explosion of the fuel tank. In this tracking camera view, plumes from one of the solid rocket boosters and other debris can be seen—as can the crew cabin. It was later reported that at least some of the seven crew survived the explosion before being killed by the high-speed impact with the sea.

04

Putting a man into orbit around the earth was a major objective of both the U.S. and the Soviet space programs. In the early Cold War arena the Soviet Union enjoyed considerable success. In the United States the Mercury astronauts were undergoing intensive training, and development of launchers and spacecraft was proceeding swiftly, if not entirely successfully. But it was perhaps inevitable that the USSR should enjoy another major coup when Yuri Gagarin was launched into space on April 12, 1961, aboard *Vostok (East) 1*. He is shown here in the capsule before launch. The mission was conservative in its aims. It was totally automated, with no cosmonaut control, and it lasted for one orbit only before Gagarin returned to Earth—by parachute in the latter stages. It was another Soviet triumph, accentuated by Gagarin's personable nature.

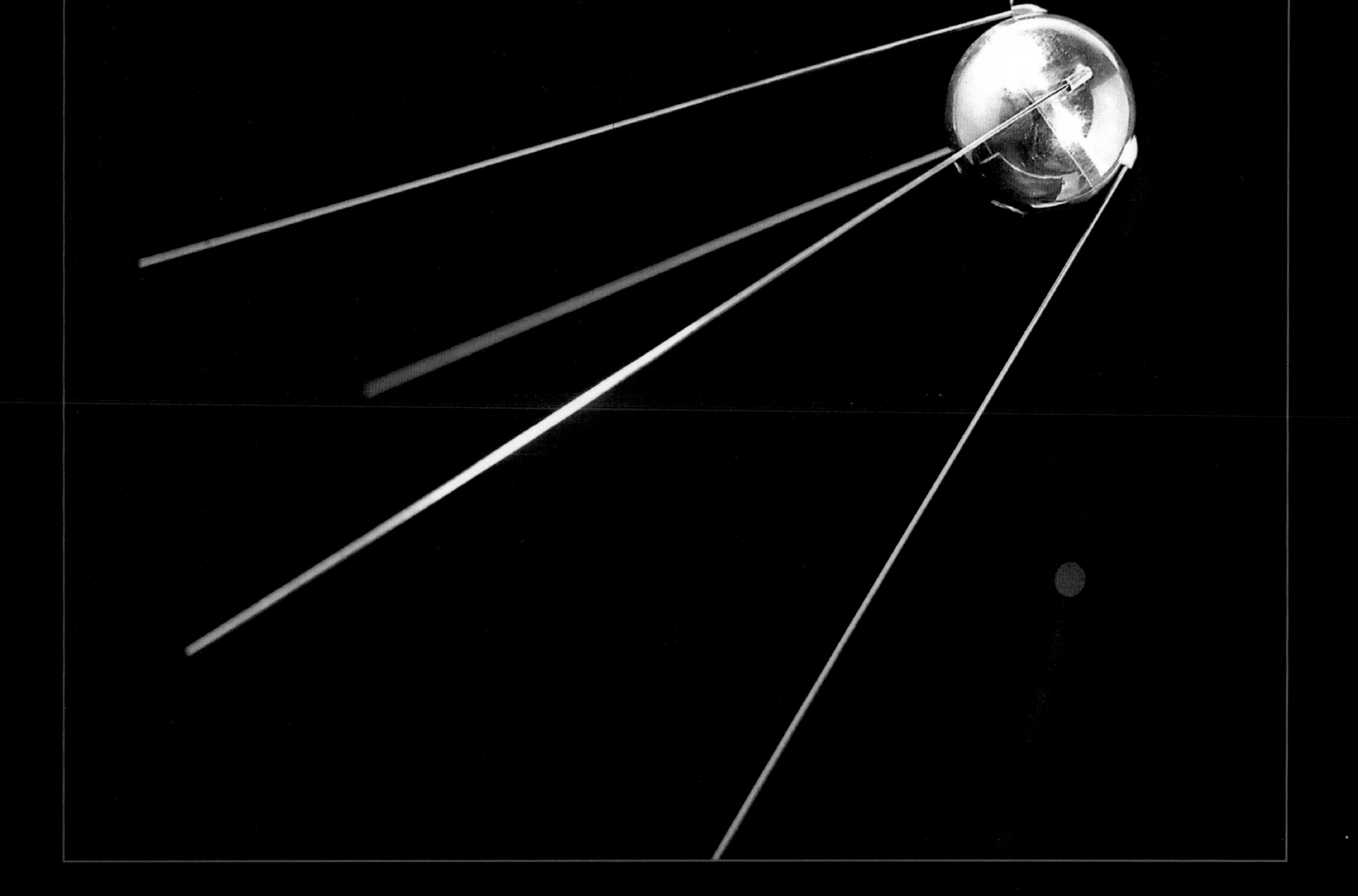

03

In the late 1950s, the Cold War between the Soviet Union and the United States was at its height. In 1957, U.S. ballistic missile development was faltering and plans for the launch of an orbiting satellite were delayed. In Russia, Communist Party General Secretary Nikita Khrushchev, after initial refusals, responded favorably to pressure from leading Soviet rocket engineer Sergei Korolev to launch a satellite aboard an SS-6 missile. The satellite was a simple aluminium sphere with four antennae, carrying a radio beacon, the signals from which could be picked up by radio hams the world over. *Sputnik 1* (sputnik is Russian for "satellite") was launched successfully at 10:28 p.m. Moscow time on October 4, 1957, and before long its characteristic "bleep, bleep" was being heard all over Western Europe and the United States. The space age had begun, and the Soviet Union enjoyed a propaganda triumph, having overtaken the United States in science and technology.

02

Since it was launched from the Space Shuttle in 1990, the Hubble Space Telescope has produced a feast of astronomical data and numberless as well as frequently beautiful images, which have captivated astronomers and general public alike. This image—dubbed the Hubble Ultra Deep Field—is a combination created from two instruments aboard the telescope and constitutes the deepest view of the visible universe ever achieved, with investigators searching for "young" galaxies with an age of between only 400 and 800 million years after the Big Bang. Many of the 10,000 galaxies recorded in the view are very different from the classical elliptical and spiral galaxies of today and are evidence of a universe that was far younger and more chaotic.

01

Edwin "Buzz" Aldrin, pilot of the Lunar Module on *Apollo 11*, took this image in July 1969. It was taken as part of a scientific investigation of the properties of the "soil" on the moon's surface. But it achieved an iconic status in the public imagination, since it represents tangible evidence of that "giant leap" for humankind in which the bounds of Earth were overcome and a human being finally set foot on another body of the solar system. Moreover, the imprint would not be quickly erased: on the moon, where there is no atmosphere, it will remain for hundreds of thousands of years, disappearing almost imperceptibly as a result of micro-meteorite bombardment from space.

Opposite: The asteroid Ida captured by the *Galileo* in 1993. Ida is approximately 32 miles (52 km) wide and has a satellite—the first natural satellite of an asteroid ever discovered. Provisionally designated "1993 (243) 1," the satellite was given the name Dactyl.

07 MINOR PLANETS

DISCOVERY OF THE ASTEROIDS

Even a casual glance at a plan of the solar system shows that the system is divided into two well-defined parts. First, there are the four rocky planets, Mercury, Venus, Earth, and Mars, with diameters ranging from 7,926 miles (12,756 km— Earth), down to only 3,030 miles (4,880 km—Mercury.) Then comes a wide gap before we reach the four giants, Jupiter, Saturn, Uranus, and Neptune. In this gap move tens of thousands of dwarf worlds known variously as minor planets, planetoids, or asteroids.

In the eighteenth century, astronomers were intrigued by the gap, which was certainly marked; Mars is on average 142 million miles (228 million km) from the sun, Jupiter as much as 483 million miles (777 million km.) Could another planet lurk there? It seemed possible, though if it existed it would certainly be small and faint. In 1800 a group of astronomers met at Lillenthal, near Bremen, where the famous observer Johann Schröter had set up and astronomical center. They formed an association, which they called the "Celestial Police," and began a systematic search for the elusive worlds. In fact, they were forestalled; in Sicily, at the Palermo Observatory, Giuseppe Piazzi was compiling a new star catalog, and on January 1, 1801—the first night of the new century—he happened upon a starlike object that moved slowly against its starry background, and which proved to be a small planet. Piazzi named it Ceres, after the patron goddess of Sicily.

However, Ceres was a real midget—a mere 600 miles (1,000 km) across—and the "Police" were not satisfied. They went on with the hunt, and soon found three more small bodies: Pallas in 1802, Juno in 1804, and Vesta in 1807. All were smaller than Ceres (although Vesta was brighter.) No more seemed forthcoming, and the "Police" disbanded, partly because Schröter's observatory was destroyed by invading French troops. The fifth asteroid was found in 1845 by a German amateur, Karl Hencke; others followed, and since 1847 no year has passed without its quota of new discoveries. When an asteroid has been tracked for long enough for its orbit to be worked out, it is given a number. Ceres, of course, is no 1; the highest numbered asteroid by June 2006 was 118172 Vorgebirge

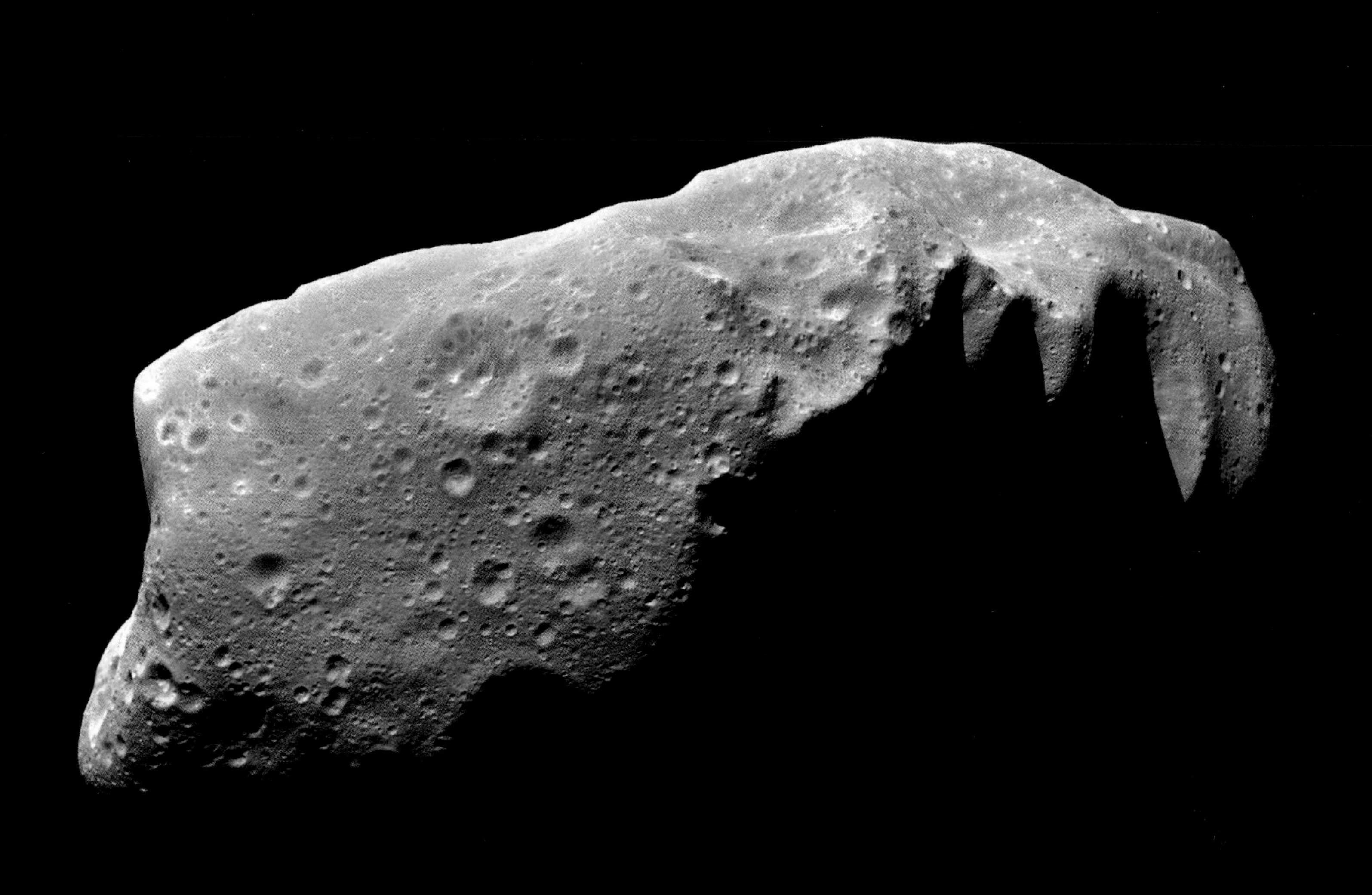

These numbers replace provisional designations. The names given to early asteroids were mythological, but the supply of gods and goddesses was soon exhausted, and some of the later names are decidedly bizarre—for instance 6042 Cheshirecat (remember Lewis Carroll?), 1581 Abanderada (someone who carries a banner), 2397 Lappajarvi (a Finnish meteorite crater, now filled by a lake), and 2309 Mr Spock (after a ginger cat that was named in honor of the Vulcanian and highly intelligent character in *Star Trek*). Mercifully, names of modern politicians and military leaders are banned by the relevant Commission of the International Astronomical Union, so there is no danger of Winston Churchill being given an orbit close to Adolf Hitler.

What do we know about the origin of the asteroids? We can now say "a good deal," because the problem is bound up with that of the origin of the solar system itself, and here we are entitled to believe that we are at least on the right track.

We are confident about the age of the earth: 4.6 thousand million years. All the planets condensed out of a cloud of dust and gas surrounding the youthful sun, which admittedly was not as powerful then as it is now. Close in, the temperatures were high, and the fledgling planets were unable to retain their lightest elements, notably hydrogen, so they evolved into small bodies rich in heavy elements such as iron. Further out the temperatures were much lower, and planets formed there were able to hold on to their hydrogen and helium, so they could grow to enormous sizes. Jupiter, the largest of all, has a diameter not far short of 90,000 miles (145,000 km). Certainly it must have a hot, silicate core, but the bulk of the globe is made up of liquid hydrogen overlaid by a hydrogen–helium "atmosphere"; look at Jupiter through a telescope and you will see the top of the cloud layer. Jupiter's mass is 318 times that of the earth; in fact its mass is greater than those of all the other planets combined. The other colossus is Saturn, which, however, is only 95 times as massive as the earth and is, of course, much further from the sun.

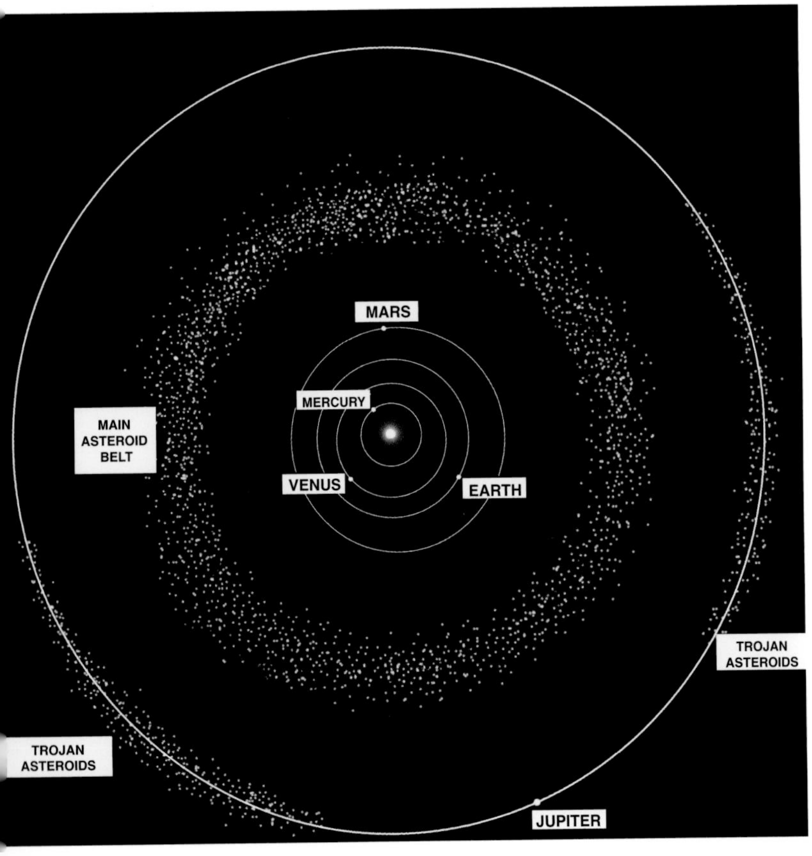

Jupiter is the key to the asteroid situation. Its gravitational pull is immense, and every time a planet began to form in the region closely outside the orbit of Mars it was disrupted by Jupiter's pull. The end product was therefore not a single large planet, but a swarm of dwarfs. There are also gaps in the main asteroid zone, again due to Jupiter, which are known as Kirkwood Gap, in honor of the American astronomer Daniel Kirkwood, who drew attention to them in 1857. There are certain regions which Jupiter keeps "swept clear."

The old theory that the asteroids represent the debris of a former planet that exploded for some reason or other has been resigned. So has the idea that the asteroids were formed as a result of a collision between two larger bodies. However, asteroid collisions must once have been common, and indeed may still be so today, as we can tell from the fact that many of the smaller members of the swarm are irregular in shape (one, 216 Kleopatra, is shaped remarkably like a dog's bone—see page 155). Also, collisions produce asteroid "families" whose members presumably have a common origin.

THE MAIN BELT

The "Main Belt" of asteroids is fairly well defined. At the inner edge we have 434 Hungaria, whose mean distance from the sun is 172 million miles (277 million km) and whose orbital period is 2.7 years. It is the brightest member of a small family. At the outer edge is the lonely 279 Thule, at a mean distance of 395 million miles (636 million km) and a period of 8.2 years. Searches for members of a Thule family have been

Left: The inner solar system, from the sun to Jupiter, showing the position of the main asteroid belt and the "Trojan" asteroids, which move in the same orbit as Jupiter.

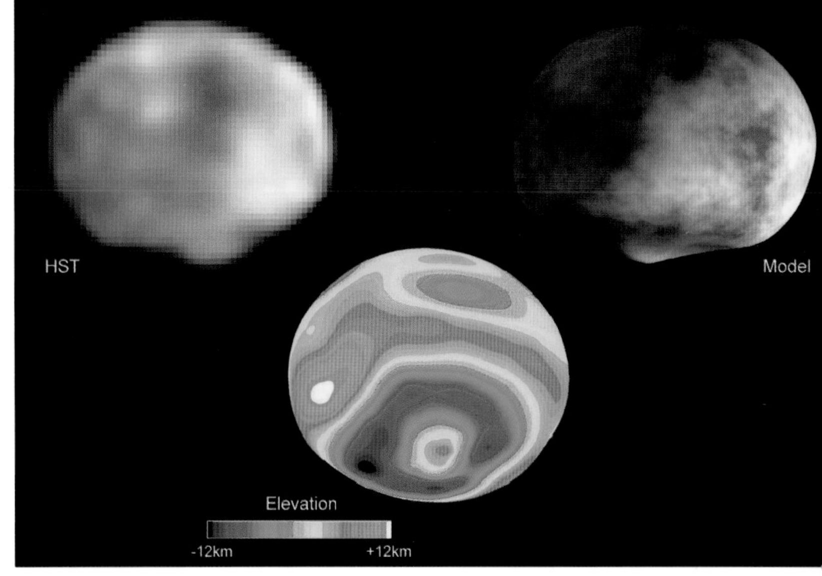

unsuccessful. Thule, with a diameter of 81 miles (130 km), is large by asteroidal standards; Hungaria is less than 7 miles (12 km) across.

Only six Main Belt asteroids are as much as 200 miles (322 km) in diameter. They are 1 Ceres (slightly oblate; longest diameter 602 miles/969 km), 2 Pallas (triaxial; longest diameter 355 miles/571 km), 4 Vesta (326 miles/525 km), 10 Hygeia (267 miles/430 km), 704 Interamnia (210 miles/338 km) and 511 Davida (201 miles/323 km). 3 Juno comes further down the list. It is safe to say that no Main Belt asteroids over 100 miles (160 km) across remain to be discovered.

Through ordinary telescopes an asteroid looks exactly like a star and betrays its nature only, because of its apparent movement against the background of real stars. To see any surface details is beyond the range of any but the world's very largest telescopes, and even then not much can be made out. The Hubble Space Telescope has mapped Vesta in some detail, but for most of our information we rely on space probes, and there have been several "close encounters" plus one actual landing—not on one of the largest bodies, but on the diminutive, sausage-shaped 433 Eros.

We can learn a great deal from examining the spectra of asteroids, and we find that they are far from alike. There are various definite types, classified according to their physical and surface characteristics. For example, there are the C-type bodies which are rather dark, and have specters like those of meteorites known as carbonaceous chondrite; Ceres is of this type. There are asteroids rich in metals (type S), such as Juno; there are some, such as 16 Psyche, made almost entirely of nickel-iron; the surface of Interamnia is clay-coated; 246 Asporina is almost pure olivine, and so there is plenty of variety. (Olivine is made up of magnesium, iron, and silicon; its chemical formula is (Mg, Fe)2Si04. There have been suggestions that in the future we might try out mining

Above: Gaspra, photographed by the *Galileo* spacecraft in 1991. A striking feature of Gaspra's surface is the abundance of small craters. More than 600 craters, 330–1650 feet (100–500 m) in diameter are visible here.

Above right: Three views of the asteroid Vesta. The top left image shows Vesta as seen by the Hubble space telescope, and top right is a computer model based on this. The bottom image is a map of the varying surface elevation of the asteroid.

operations on asteroids and bring home valuable materials, but this does sound more like science fiction than serious forecasting.

It cannot be said that the asteroids have always been popular among astronomers. Plates exposed for quite different reasons were often found to contain dozens of asteroid tracks, all of which had to be identified, which wasted an incredible amount of time. One irritated German even referred to them scathingly as "vermin of the skies," and a hundred years ago a well known astronomical author, G. F. Chambers, wrote that "the minor planets are of no interest to the amateur; in fact they are of very little interest to anybody." Opinion today is very different from what it was in the past. Asteroids are closely studied, and indeed one major observatory, the Klet Observatory in the Czech Republic, is devoted entirely to them. They are important—and some of them may well be dangerous!

Ceres is the largest of the Main Belt asteroids. It is spherical, or nearly so, and is now believed to have a dense core surrounded by a mantle rich in water ice. The escape velocity is only 0.4 miles (0.6 km) per second, but there does seem to be an excessively thin atmosphere, while the gravitational pulls of the other Main Belt asteroids are too weak for them to retain any atmosphere at all. Images taken with the Hubble Space Telescope have shown a large dark spot on Ceres, which may well be a crater; it has been named "Piazzia." If it really is an impact structure, the impactor must have been at least 15 miles (24 km) across. The rotation period is just over 9 hours—a normal rate for a large asteroid, although small ones may spin much more quickly. The tiny 2001 CF84 has a period of only 29 minutes; it is so small—around ½ mile (0.8 km) in diameter—and has so feeble a pull of gravity that any loose object on the surface would be promptly thrown off. Landing there would be quite a problem!

Vesta is smaller than Ceres, but looks brighter, because it is closer in and has a more reflective surface. Geologically it is very interesting. There are ancient lava flows and basaltic regions, so that Vesta must once have had a molten core. There are two distinct types of terrain; one hemisphere shows quenched lava flows, while the other shows characteristics of old molten rock which cooled and solidified underground. There is one large impact crater, with a central peak. Vesta has an unusual type of spectrum, and it has been suggested that two smaller asteroids of the same type, 9969 Braille and 1929 Kollua, may be broken off fragments of it.

CLOSE ENCOUNTERS

Several Main Belt asteroids have been surveyed by passing spacecraft. The first was 951 Gaspra, on November 29, 1991; the *Galileo* probe, en route for Jupiter, imaged it from a range of 10,000 miles (16,000 km), and found it to be wedge-shaped, like a distorted potato. It measures 12 x 7.5 x 7 miles (19 x 12 x 11 km), with a darkish, cratered surface. The probe also imaged 243 Ida, which is accompanied by a very small satellite, now named Dactyl. 253 Mathilde was passed by a different spacecraft, *NEAR-Shoemaker*, and was a surprise; it is as black as charcoal, and although it is only 41 miles (66 km) across it has a rotation period of over 17 days. There are huge craters, and it has been said that Mathilde was porous. Fittingly, the main craters have been named after famous coalmines.

The spacecraft that imaged Mathilde was originally called *Near Earth Asteroid Rendevous*; the name was added later in honor of the American astronomer/geologist Eugene Shoemaker. Shoemaker never went to the moon as he had so badly wanted to do during his lifetime, but at least his ashes were scattered there by the *Apollo* astronauts.

Opposite: Eros. The arrow shows the landing site of the *NEAR-Shoemaker* satellite on February 12, 2001, just to the south of the saddle-shaped feature Himeros. To the surprise of the controllers, the craft was undamaged and transmitted for another month.

Above: The largest crater on Eros, known as Psyche. The crater is 3.3 miles (5.3 km) wide and is surrounded on one side by smaller sister craters, making it look rather like a giant paw print.

Below left: Itokawa, discovered in 1998. The Japanese probe *Hayabusa* landed on its surface in 2005, its mission to take soil samples. It won't be clear until 2010 whether it was successful, however.

Below right: Mathilde, imaged by *NEAR-Shoemaker*. The asteroid has vast craters, particularly the one at the center, which is estimated to be more than 6 miles (10 km) deep.

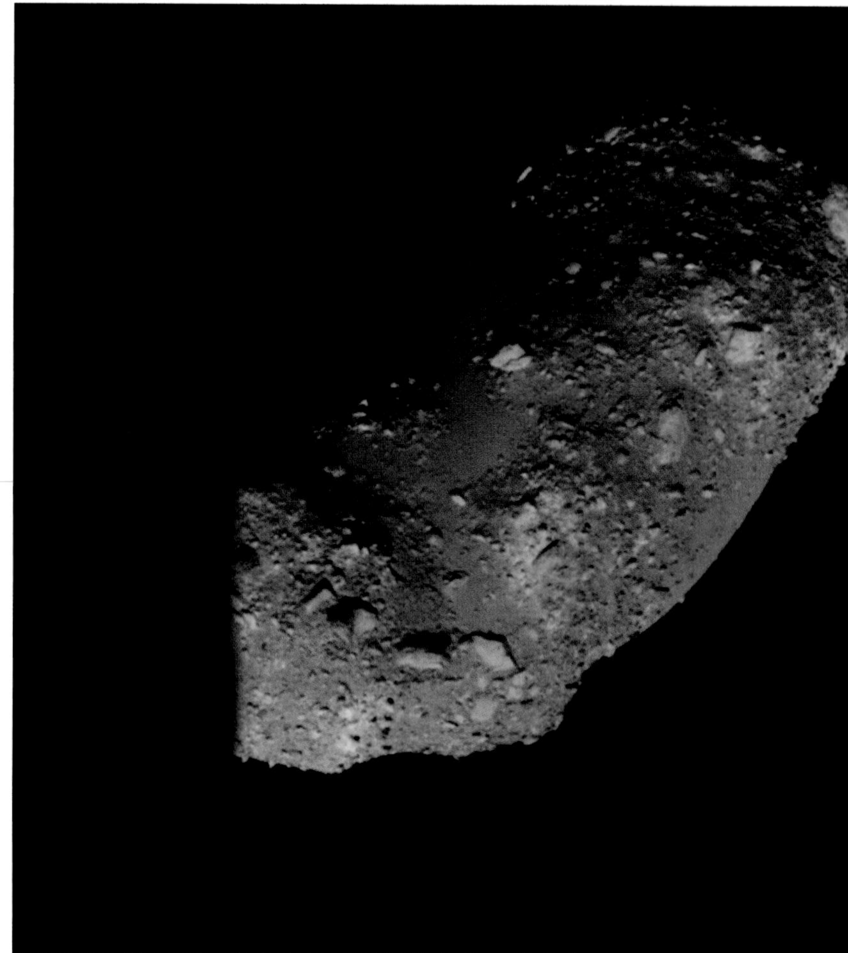

Before discussing the asteroid landing, it seems necessary to say something about asteroids which do not remain in the Main Belt. Eros, discovered in 1898, was the first to be found. It stays in the Main Belt for most of its 1.76-year period, but at perihelion it comes within the orbit of Mars, and may pass within 16 million miles (26 million km) of the earth, as it did in 1936. The last close approach was in 1975. Since then other asteroids moving closer-in than the Main Belt have been found. Some, like Eros, have orbits which cross that of Mars, but not the earth; others cross the path of both Mars and Earth, a few even swing closer in than Mercury. 1566 Icarus is one of these. Its orbital period is 106 days, and at perihelion it is a mere 17 million miles (27 million km) from the sun so that its surface is red hot; at aphelion it retreats to over 165 million miles (266 million km), the bitterly cold region well beyond Mars. Icarus-type asteroids must have the most uncomfortable climates in the whole solar system. However, all close-approach asteroids are tiny; only one, 1036 Ganymede (diameter 26 miles/42 km) is larger than Eros.

NEAR-Shoemaker was launched from Cape Canaveral on February 19, 1996. On June 27, 1997, it sent back images of the black "coal mine" asteroid 253 Mathilde, and in February 2000 it went into orbit round Eros to begin a year-long study. Eros, of course, has so low an

Below: Kleopatra's remarkable dog-bone shape and unusual composition—its surface is loosely consolidated rubble, possibly with a solid core—may result from two similarly sized asteroids colliding and sticking together instead of breaking apart.

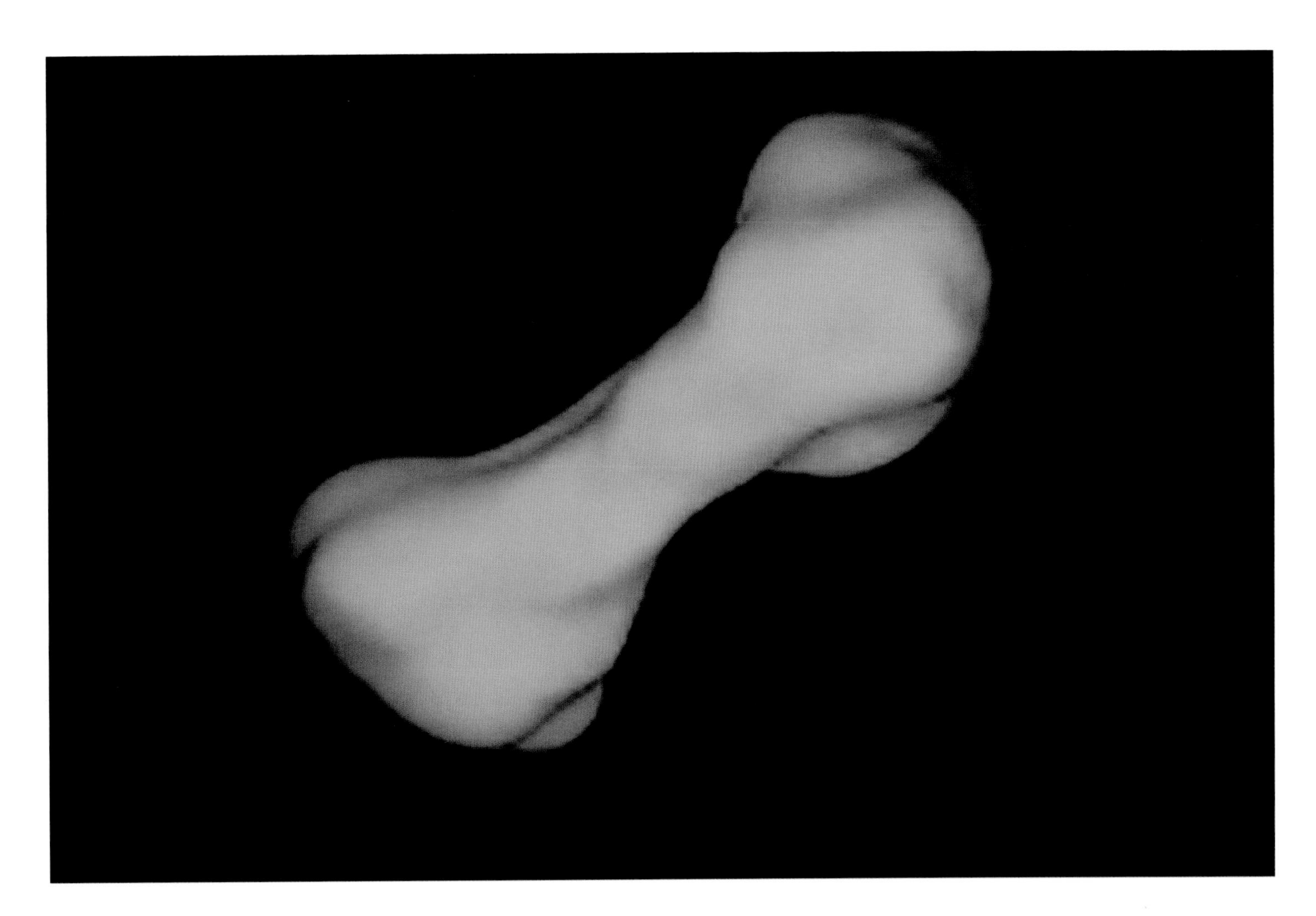

escape velocity (32 feet/9.8 meters per second) that it cannot hold on to even a trace of atmosphere. The daytime temperature was 212°F (100°C), but this fell to to –148°F (–100°C) at night. As Eros has a short rotation period (just over 5 hours) the surface temperature changes very quickly.

The surface was rugged, as was expected, and there were many craters between 1,600 and 3,000 feet (500 and 1,000 m) across, some with flat floors ("ponds"). The craters were named after characters associated with romance, such as Cupid, Psyche, Lolita, and Don Quixote. Eros seemed to be a very primitive body, and also very ancient.

After the orbital survey, *NEAR-Shoemaker* still had some fuel left, and the planners at NASA decided to attempt a controlled landing. Against all the odds they succeeded, and on February 12, 2001, the probe touched down gently in the middle of a saddle-shaped feature which had been christened Himeros after the Greek god of desire—always represented as a winged youth. The main instruments went on working after the impact, and contact was not finally lost until February 28. *NEAR* remains on the surface of Himeros, waiting to be collected and put on display at the nearest cosmic museum.

The second (and so far, the last) short landing was by a Japanese probe, *Hayabusa*, on the asteroid 25143 Itokawa, which is irregular in shape. It is a real midget, only 2,300 feet (700 m) long with a maximum width of 1,000 feet (300 m). It has an orbital period of 550 days, and a rotation period of 12 hours. At perihelion it just crosses the earth's orbit.

The Japanese plan was to touch down on the surface, collect samples of material, and bring them home. *Hayabusa*, also carried a very small box-shaped vehicle, Minerva, which was to be released to make an independent landing after which it would "hop" along and transmit pictures. Unfortunately, there were serious problems almost from the outset. *Hayabusa* was launched on May 9, 2003 and reached the neighborhood of Itokawa on the following September 16, going into an orbit round the asteroid and sending back good images. Minerva was released on November 14, but missed its target and sadly drifted away into space. The main vehicle did touch down, but apparently was unable to collect its samples; a second attempt was made before the start of the return journey. Whether any material was collected we will not know until *Hayabusa* returns—if indeed it does come back safely. If all goes well, the entry probe will be detached and will eventually parachute down through the atmosphere. It is scheduled to land in Australia in 2010.

Itokawa was a surprise. Far from being a solid rock, it was better described as "a heap of rubble." The surface was littered with boulders and gravel, and its density is so low that the whole body may be porous, the material is very loosely packed, and is barely held together by the asteroid's week gravitational pull. There are several smoother areas, and *Hayabusa* came down in one of these, which the Japanese named the Moses Sea—the probe was originally known as Muses-C.

EARTH COLLISIONS

Itokawa is not moving in a path that can bring it into collision with the earth, but this is not true of other small asteroids which invade the inner part of the solar system. There may well have been major impacts in the past. It is widely believed that a large object hit the earth 65,000 years ago, causing a huge crater in the region of the Yucatán peninsula in Mexico; the effects were widespread, and changed the whole climate, with disastrous results for the dinosaurs, which could not adapt to the new conditions and died out. This may or may not be true, but if it were, for example, a kilometer-wide asteroid the results would be catastrophic. We would probably cope better than the dinosaurs did, but there would be thousands of millions of casualties. The danger from wandering asteroids may be slight, but it is not negligible.

Several asteroids have been shown to pass between the earth and the moon, but the most menacing asteroid so far to be discovered is 99942 Apophis, discovered on June 19, 2004. It may be only just over 1,000 ft. (305 m) across, but several times during the next 30 years it will come close enough to cause a certain amount of alarm; it will indeed skim the uppermost part of the atmosphere. The first pass will

Above: Hubble images of Pluto revealed two previously unknown moons: Nix and Hydra. These moons are approximately 5,000 times fainter than Pluto and are about two to three times farther from it than its large moon, Charon.

be in 2029, but there will be another on April 13, 2036, and Apophis will then be easily visible with the naked eye. The odds of an impact are thousands to one against, but there will certainly be a great deal of popular interest!

If we detect an Apophis-style asteroid on a collision course, is there anything that we can do about it? Possibly, yes—provided that we find it in time. Shattering it with a nuclear missile would be pointless, even if it could be achieved, the object would simply be changed into a shower of shrapnel. The only solution would be to divert it—in fact, "nudge" it into a path that would make it avoid the earth. An atom-bomb exploded very close to it might suffice, but the operation would be extremely difficult.

OUTLIERS

Just as there are asteroids that swing in from the Main Belt, so there are asteroids which are much further out. First there are the "Trojans," which move in the same orbit as Jupiter (they are called Trojans because the brightest of them have been named after warriors of the Trojan War). The first to be found was 588 Achilles, by Max Wolf in 1906; by now hundreds are known, but all are faint, mainly because they are so far away. They are not swallowed up by Jupiter because they keep prudently well ahead of or well behind the Giant Planet. Mars has several Trojans, and so has Neptune; probably there are Trojans also associated with Saturn and Uranus.

There are asteroid-sized bodies which move in the far reaches of the solar system—for example 2060 Chiron, whose path lies between those of Saturn and Uranus. There are asteroids with very eccentric orbits, taking them from the neighborhood of the earth out far beyond the Main Belt. Now we have located a new swarm, which is known as the Kuiper Belt in honor of the Dutch astronomer Gerard Kuiper, who discovered it as long ago as 1951—before any of its members had been found.

The outermost of the giant planets, Neptune, moves around the sun at a mean distance of 2,793 million miles (4,495 million km). In 1930 Clyde Tombaugh, working at the Lowell Observatory in Arizona, discovered Pluto, which was naturally accepted as being the ninth planet. It was small, and had an orbit that was both eccentric and inclined; at perihelion it actually moves closer in than Neptune, even though for most of its 248-year period it is much further out.

Pluto is of the 14th Magnitude, and can therefore be seen as a starlike point in a fairly small telescope, but it was always enigmatical; it did not seem to fit in to the overall scheme of the sun's family. It was found to have a very thin atmosphere, but when it moves in the far part of its orbit it becomes so cold that the atmosphere must freeze out on to the surface. Vague dark patches and brighter regions could be imaged with the Hubble Space Telescope, and it found that the rotation period is 6 days 9 hours.

In 1978, James Christy, at the U.S. Naval Observatory at Flagstaff, found that Pluto has a satellite, Charon, whose diameter is more than half that of Pluto itself: 750 miles (1,200 km) against Pluto's 1,440 miles (2,320 km). The two are only 10,000 miles (16,000 km) apart, and Charon has an orbital period of 6.4 days—the same as Pluto's rotational period; to anyone standing on Pluto, Charon would stay motionless in the sky. Two more attendants, Nix and Hydra, were found in 2006, but both were very small.

Peculiar though it was, Pluto and its satellites seemed to be the only occupants of the desolate region beyond Neptune. However, in 1992, David Jewitt and Jane Luu, from Palomar, found a small body—120 miles (190 km) across, which proved to be the first of the Kuiper Belt Objects. Others were soon added, and by now it is clear that hundreds, perhaps thousands, exist. Several are larger than Pluto, and have satellites of their own. There is no reason why an even larger object should not exist.

All Kuiper Belt Objects (KBOs) are faint because they are so remote. What does this mean with regard to Pluto? We have to admit that if Pluto is classed as a true planet, then so are many other KBOs, and this makes no sense. We now know that Pluto is not exceptional, and we have to relegate it to the state of an ordinary KBO.

Apart from Pluto, the KBOs are too faint to be imaged with instruments available to most amateur observers, and we do not know much about their physical characteristics. Again we must turn to space research methods, and when the New Horizons probe arrives at its target, hopefully within the next ten years, we may at last start to understand these far-away and intriguing members of the sun's family.

Opposite: True-color mosaic of Jupiter, constructed from images taken by the narrow-angle camera on Cassini during the pass made on December 29, 2000, from a range of approximately 6,200,000 miles (9,977,933 km). The Great Red Spot is very evident. The colors shown here are much as would be seen by a human observer at the same range.

08 GIANT PLANETS

The four giant planets move in the vast gap between the two zones of small bodies, the Main Belt asteroids and the Kuiper Belt objects. These two zones are very different. All the Main Belt asteroids combined would not make up one body as massive as the earth, and Ceres, the largest of them, is a mere 600 miles (966 km) in diameter. The Kuiper Belt contains many bodies larger than Ceres, and it has been estimated that their total mass is at least 30 times that of the earth, so they have hammered effects on the whole of the outer part of the sun's family. Note, too, that the outer planets are not alike. Jupiter and Saturn are "gas-giants," while Uranus and Neptune are better described as "ice-giants."

JUPITER

Jupiter is the dominating force, and it has even been said that the solar system is made up of the sun, Jupiter, and assorted debris. In our skies, Jupiter shines more brilliantly than any other planet apart from Venus and (very occasionally) Mars; it comes to opposition approximately every 13 months, and this means that it is well placed for observation for several months in every year. It is a fascinating object, if only because it is always changing; one never knows what will happen next.

Jupiter has a core, presumably silicate, at a high temperature; 36,000°F (20,000°C) is one estimate, but may be too cool. Jupiter sends out more energy than it would do if it depended entirely upon what it receives from the sun. On the other hand, it is misleading to regard Jupiter as a "failed star." A normal star, such as the sun, shines because of nuclear reactions going on deep inside it; a temperature of around 18,000,032°F (10,000,000°C) is needed to trigger these reactions, so Jupiter falls short by a very wide margin. It would have to be several times as massive as it actually is.

Surrounding Jupiter's core are layers of liquid hydrogen. The lower parts are so compressed that they take on the characteristics of a metal; the upper layers are molecular. Above comes the dense "atmosphere," the top of

which we see when we look at Jupiter through a telescope. Hydrogen is the main constituent, making up around 80 percent of the total; helium accounts for 14 percent which does not leave much room for anything else. There are hydrogen compounds, notably methane, ammonia and hydrogen sulphide; there is certainly less than 1 percent of water. Gases warmed by the internal heat of the planet rise into the upper atmosphere and cool, producing high-altitude clouds of ammonia crystals. These clouds form the bright zones on Jupiter, which are both higher and colder than the dark belts. The overall temperature of the visible gas layer is around –150°F (–100°C).

Telescopically the yellowish disc is striped with the dark belts. Any small telescope will show several of these. The two most prominent are the two Equatorial Belts lying either side of the equator. It is easy to see that the disc is markedly flattened, and indeed the equatorial diameter is over 5,000 miles (8,000 km) greater than the diameter as measured through the poles. This is because, as Jupiter is such a quick spinner, the equatorial zone bulges out. A Jovian "day" is less than ten hours long, but the planet does not rotate in the way that a rigid body would do; the rotation period between the edges of the equatorial belts is 9 hours 30 minutes (System 1) while over the rest of the planet the period is five minutes longer (System 2). There is, in fact, a powerful equatorial jet stream, and, moreover, some discrete features have periods of their own, so that they drift around in longitudes but not in latitudes.

As well as the belts there are spots, "wisps," and festoons. Of special note is the Great Red Spot, which intrudes into the belt known as the South Temperate. It was once thought to be a floating mass, but is now known to be a whirling storm—a phenomenon of Jovian meteorology; its surface area is greater than that of the earth. It can sometimes appear very red, although the cause of this color is not known—phosphorus has been suggested. In much the same latitude are bright "white ovals," and in 2006 one of these suddenly became decidedly red. Whether this hue will persist remains to be seen.

The first spacecraft to be sent to Jupiter was *Pioneer 10*, which flew past the planet in December 1973; its twin, *Pioneer 11*, followed a year later, and then moved on to a rendezvous with Saturn. Both *Pioneer*s were very successful, and sent back images of both Jupiter and its major satellites; they also confirmed that there is an intensely powerful magnetic field, together with zones of lethal radiation which would quickly kill any astronaut unwise enough to enter them. Then in 1977 came *Voyagers 1* and *2*, which improved on the *Pioneer* results and discovered a system of thin, dark rings, quite unlike the glorious rings of Saturn and far beyond the range of ordinary Earth based telescopes. *Galileo*, more ambitious still, was sent up on October 18, 1989, but before it reached Jupiter, in December 1996, astronomers had been privileged enough to witness a truly remarkable event: a collision between a planet and a comet.

In March 1993 a team of American observers—Eugene and Carolyn Shoemaker, and David Levy—discovered what they described as "a squished comet"; it was their ninth success, and so the comet was listed as Shoemaker-Levy 9 (SL9 for short). It was found to be on a collision course with Jupiter; the planet's strong gravitational pull had torn the comet apart long before impact, in July 1994, and the

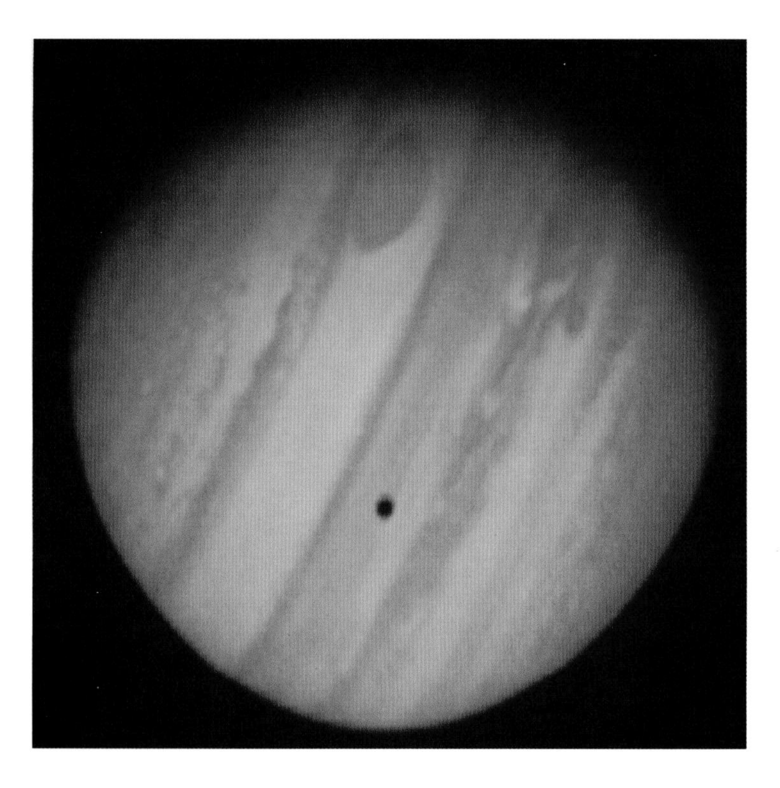

Above: Pioneer 10, which lifted off from Kennedy Space center in March 1972, was the first spacecraft to penetrate the asteroid belt and travel to the outer solar system. The spacecraft passed within 81,000 miles (130,357 km) of Jupiter in December 1973, obtaining the first close-up images of the planet and its satellite Io (seen here as a small black dot).

Opposite: *Galileo* begins its journey on October 18, 1989, when it leaves the cargo bay of the space shuttle *Atlantis*. It began its free-fall flight to Venus for the first of three gravity-assist flybys which would take it to Jupiter. From the shuttle, Commander Donald Williams declared: "*Galileo* is on its way to another world. Fly safely!"

USA

Left: Impact site of one of the fragments of the comet Shoemaker-Levy 9, July 18, 1994. This image was taken by the Hubble Space Telescope 1 h 45 m after the actual impact. The site is the dark round area below and left of the center of the frame, at latitude 45 degrees South. The impact velocity was 37 miles per second (59.5 km per second).

fragments cascaded into the Jovian clouds, causing disturbances visible from Earth with almost any small telescope. The best views, however, were obtained by the Hubble Space Telescope and by the *Galileo* probe, which was already on its way. There were no permanent effects on Jupiter, but the disturbances were traced for over a year. Had missiles of this size hit Earth the results would have been catastrophic.

Galileo reached its target 18 months later. It carried an entry probe, which plunged into the clouds and transmitted for over an hour before losing contact; by then it had reached a depth of 370 miles (600 km) below the cloud tops, and sent back the first reliable information about conditions inside the globe. There was one major surprise; the atmosphere was much drier than had been expected—it had been thought that there would be obvious signs of water vapor. It may be that the probe entered a particularly dry region.

After releasing its entry probe, *Galileo* went into orbit around Jupiter, and began a lengthy survey, not only of the planet, but also of its satellites. Jupiter has a wealth of moons—over 60 of them—but only four are large; these are Io, Europa, Ganymede, and Callisto, known collectively as the Galileans because they were carefully studied by Galileo, the first real telescopic observer, in 1610. Europa is slightly smaller than our moon, Io slightly larger, and Ganymede and Callisto much larger—in fact, Ganymede, 3,270 miles (5,260 km) in diameter, is larger than the planet Mercury, though not so massive.

The Galileans are not alike. Ganymede and Callisto are icy and cratered, and apparently inert; Europa is also icy, but has a cracked surface looking rather as though it is made up of icebergs floating on an ocean. In fact, it is widely believed that an ocean of ordinary water does exist below the ice-layer, but even if this is so, we can hardly believe that any life could survive in such a soulless, pitch-black sea. Io, by contrast, is wildly volcanic, with massive eruptions going on all the time; the volcanoes were first seen on *Voyager* pictures, and striking images have been obtained of them. They are fiercely hot, even though the general surface temperature is below –220°F (–140°C). Io's surface is dominated by sulphur, which accounts for its yellowish color, and there is an excessively thin atmosphere of sulphur dioxide. As the satellite moves within Jupiter's deadly radiation zones, it must be just about the most uninviting world in the entire solar system.

SATURN

Saturn is also a true gas giant, with a silicate core, layers of liquid hydrogen and a hydrogen–helium atmosphere. It is smaller and less massive than Jupiter, as well as being much further away, but it still appears as bright as most of the stars, and, like Jupiter, it is well seen for several months in every year. Its orbital period is 292 years, and its rotation period is 10¾ hours, so its yellowish globe is markedly flattened. There are belts of the same type as Jupiter's but much less prominent; there is no feature comparable with the Great Red Spot, but there are occasional bright white spots, due to the uprush of material from below. One particularly brilliant spot was discovered in 1933 by a well-known amateur astronomer, W. T. Hay (remembered by many people as Will Hay, the British stage and screen comedian). It became a very easy telescopic object, but did not last for long.

Of course, Saturn is distinguished by its glorious rings, made up of ice crystals. They are very extensive, with a maximum spread of 169,000 miles (272,000 km), but they are less than a mile thick, so when they are visible side-on from the earth (as will next happen in 2009) they almost disappear. Binoculars will never show that; with a telescope, Saturn is arguably the loveliest object in the sky, and is unlike any other; the rings of the other giant planets are dark and obscure.

Under favorable conditions, ordinary telescopes show three rings round Saturn. Two rings (A and B) are bright, and are separated by a gap known as Cassini's Division in honor of the Italian astronomer who discovered it in 1675. Closer to Saturn lies Ring C, which is known as the Crêpe or Dusky Ring because it is more or less transparent. There is another gap in Ring A, known as the Encke Division, and material scattered between the Crêpe Ring and the cloud tops is often (misleadingly) referred to as Ring D.

Below: *Galileo's* first view of Ganymede—a natural color image taken in 1996. The darker areas are older, more heavily cratered regions. The brownish-gray color is due to a mixture of rocky material and ice.

Below: Another *Galileo* image, this time of Callisto. Like that of Ganymede, this image reveals ice through its cratered surface.

Below right: Europa is icy, like Ganymede and Callisto, but most traces of craters on its surface have been erased by recent geological activity.

Bottom: An active volcano on Io, as seen from *Galileo*. Io is the most volcanic world in the solar system; eruptions are going on all the time, so there is constant re-surfacing and few impact craters. As Io moves within Jupiter's lethal radiation belts, it is a world to be viewed from a respectful distance!

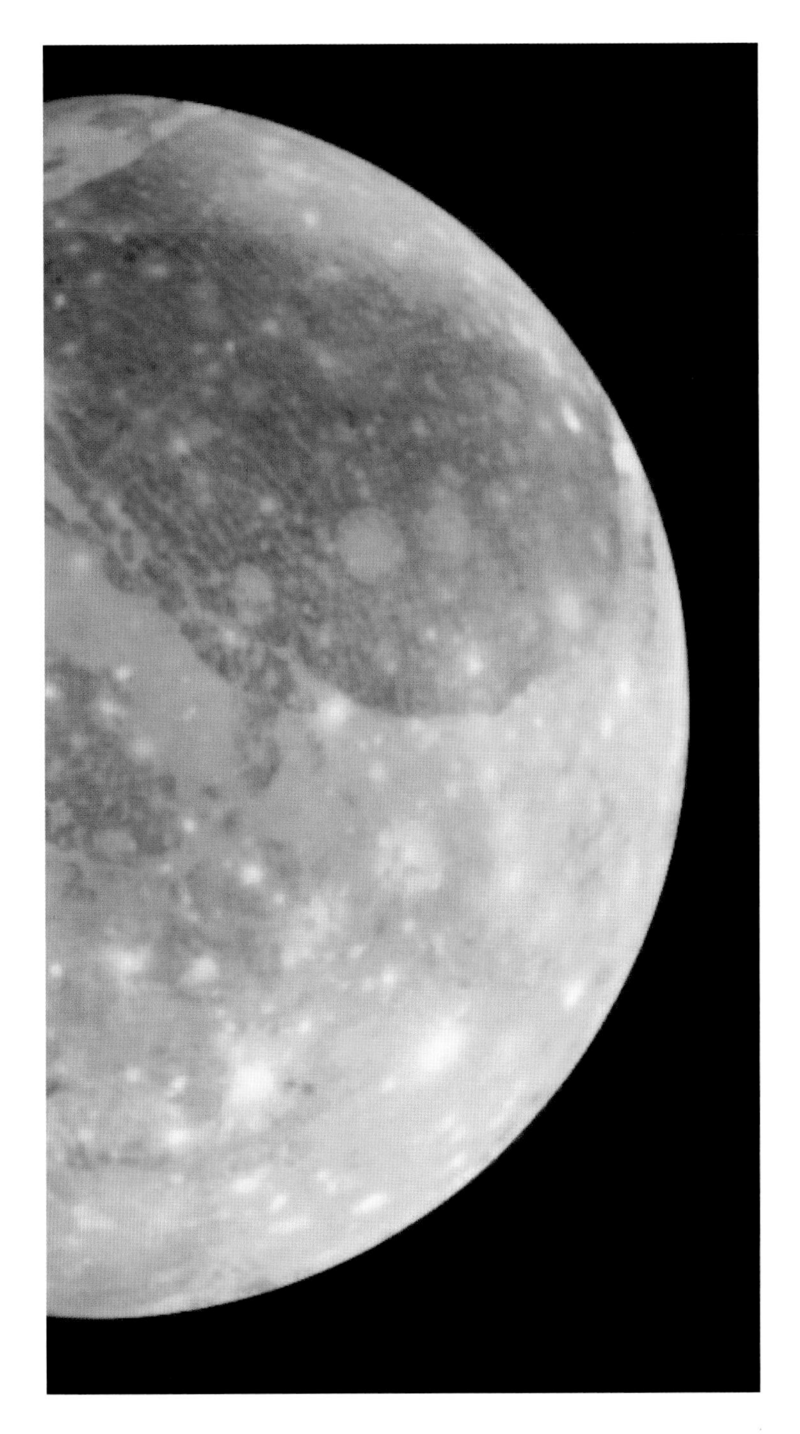

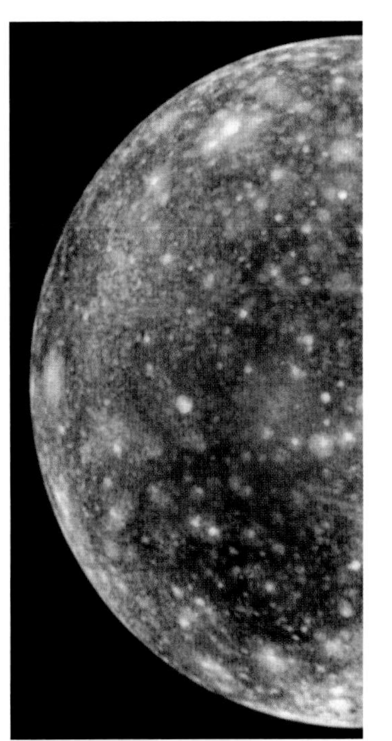

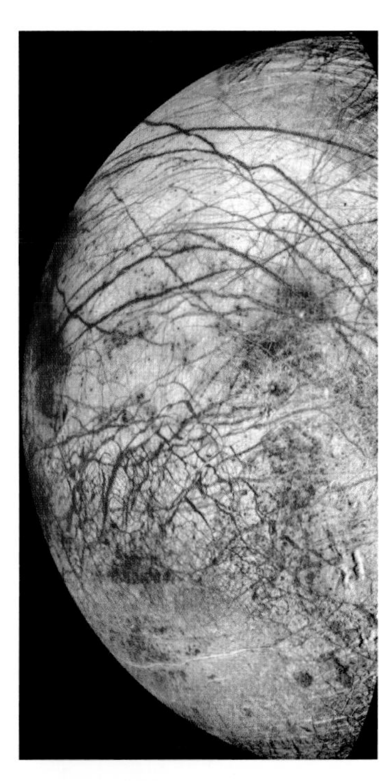

Above: Panoramic view of Saturn, created from 165 images taken with *Cassini's* wide-angle camera on September 15, 2006. *Cassini* is in Saturn's shadow, and the rings are seen as never before. The well-defined G-ring is seen outside the main system; farther out, encircling the system, is the less well-marked extended E-ring. The colors are slightly exaggerated.

MISSIONS TO SATURN

Before the space age began it was assumed that Saturn's rings must be smooth and structureless, but the first probes to reach the planet, *Pioneer 11* (1979) and the two *Voyagers* (1980 and 1981) showed that this was far from the case. There are many hundreds of narrow rings and minor divisions; it has been said that the ring system has more grooves than a gramophone record! The gravitational effects of the inner moons, mainly Mimas and Enceladus, must be responsible, but the first images sent back from close range took astronomers by surprise. There were also strange, dark, short-lived "spokes" across ring B, thought to be due to particles elevated away from the ring-plane by magnetic and electrostatic forces.

Several new rings were found outside the main system; E, F, and G. The irregular E-ring has been described as a convoluted tangle or narrow strands; it is stabilized by two small satellites, Prometheus and Pandora, which orbit to either side of the ring and prevent the particles from straying—they are known as "shepherd satellites." Next come the highly tenuous rings G and F, which are virtually invisible from Earth.

Both *Voyagers* were outstandingly successful. The first had no more encounters, but *Voyager 2* went on to rendezvous with Uranus in 1986 and Neptune in 1989.

The latest probe, named after *Cassini*, was launched from Cape Canaveral on October 15, 1997, but traveled in a rather circuitous route, and did not reach Saturn until 2004. It confirmed all the earlier results, and, unlike the *Voyagers*, was an orbiter; it went into a path around Saturn, and as is still, at the time of writing (2007) sending back amazingly detailed images. Also, it carried a small landing vehicle, *Huygens*, which touched down gently upon the surface of Saturn's main satellite, Titan.

More moons were known before 1900, but of these only Titan was large. Four more (Iapetus, Rhea, Dione, and Tethys) were above 600 miles (1,000 km) in diameter, while Enceladus, Mimas, Hyperion, and Phoebe were smaller; Phoebe was 8 million miles (13 million kilometers) from Saturn and had retrograde motion so that it was assumed to be a captured body rather than a bona fide satellite. The system was obviously different from that of Jupiter, but it proved to be no less interesting and in some ways was even more surprising.

Left: Rings of Saturn. *Voyager 2*, from a range of 1,700,000 miles (2,735,885 km). The view focuses on the C-ring, and was compiled from three separate images taken through ultra-violet, clear, and green filters. More than 60 ringlets are shown (ignore the small dark squares, which are instrumental effects). The B-ring appears to the top left.

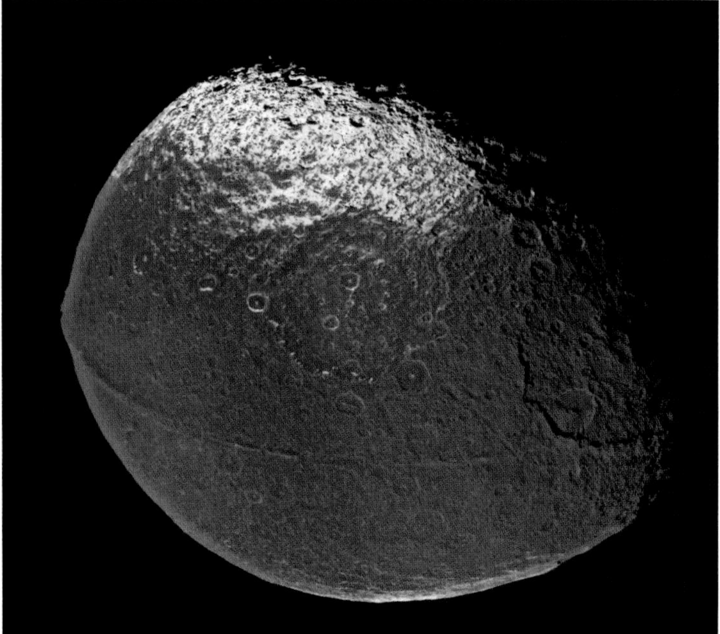

Titan, discovered by the Dutch astronomer Christiaan Huygens in 1655, was slightly larger than Mercury, and almost equal to Ganymede in Jupiter's system. It was the only satellite known to have a substantial atmosphere, discovered by Gerard Kuiper as long ago as 1944. The atmosphere hid the surface so completely that the *Voyager*s could make out nothing apart from the top of a layer described as "orange smog." When *Cassini* was safely in orbit, the *Huygens* lander was released, and—against all the odds—not only made a controlled landing, but sent back high-quality pictures. The atmosphere was found to be composed of 93.4 percent nitrogen, with mere traces of other elements, while the surface temperature was a forbidding –290°F (–179°C).

Huygens landed in a bright region, now called Adiri, and showed an extraordinary landscape; there were hills, dark areas, and what seemed to be riverbeds, although the liquid flowing through them was certainly methane rather than water. Methane rain must fall frequently, even though *Huygens* arrived in a "dry" period. It is thought that there may be methane lakes, and certainly there are wide areas of sand dunes, formed by winds generated by tidal forces from Saturn on Titan's atmosphere, which gives a ground pressure considerably greater than that of the Earth's air at sea level.

The other satellites are almost equally peculiar in their various ways. All have icy surfaces; Dione and Tethys are cratered, and Mimas has one huge crater which makes it look rather like the Death Star in the classic film *Star Wars*. Iapetus has a dark area covering almost the whole of one hemisphere, and a mountain ridge running along the equator; Enceladus gushes water vapor from its south polar region and this may indicate that, as with Europa, there may be an ocean not far below the surface. Nothing of the sort had been expected on a world less than 400 miles (650 km) in diameter, and the fountains of Enceladus have been a revelation. Remember, Saturn is so far away that from Earth the satellites look almost star-like, with only Titan showing a reasonable-sized disc.

THE ICE GIANTS

Almost all our detailed knowledge of the two ice giants, Uranus and Neptune, comes from a single probe, *Voyager 2*. Previously, we had found out comparatively little, though it was clear that they differed from the Jupiter–Saturn pair and from each other. Each has less than half the diameter of Saturn; Uranus is slightly the larger of the two, but less massive (14 times the mass of the earth, as against 17 for Neptune). Uranus is just visible with the naked eye, Neptune much too faint; telescopically Uranus is greenish, Neptune blue. Uranus takes 84 years to complete one journey round the sun, Neptune 164.9 years; both have short axial rotation periods, 17 h 14 m and 16 h 11 m respectively.

Opposite middle: Saturn's moon Iapetus, captured by *Cassini* on New Year's Eve, 2004. An amazing feature is the ridge that runs almost exactly along its equator. The ridge is approximately 12 miles (20 km) wide, and its peak is at least 8 miles (13 km) high. We do not know yet whether the ridge is formed by a mountain belt folding upwards or material erupting through a crack in the surface.

Opposite bottom: Mimas, the innermost of Saturn's main satellites, imaged from the narrow-angle on *Cassini* on August 2, 2005, from a range of 142,500 miles (229,332 km). Mimas is cratered and inert, the scene is dominated by one huge crater with a central peak, so large that the impact producing it must have been close to breaking up the whole satellite. The crater is named after William Herschel, who discovered Mimas in 1787.

Below: Lakes of liquid methane on Saturn's moon Titan, captured by *Cassini*'s radar on July 22, 2006. The existence of these lakes on Titan had been predicted more than 20 years before, but until *Cassini*'s fly-by the moon's dense haze had made it impossible to confirm.

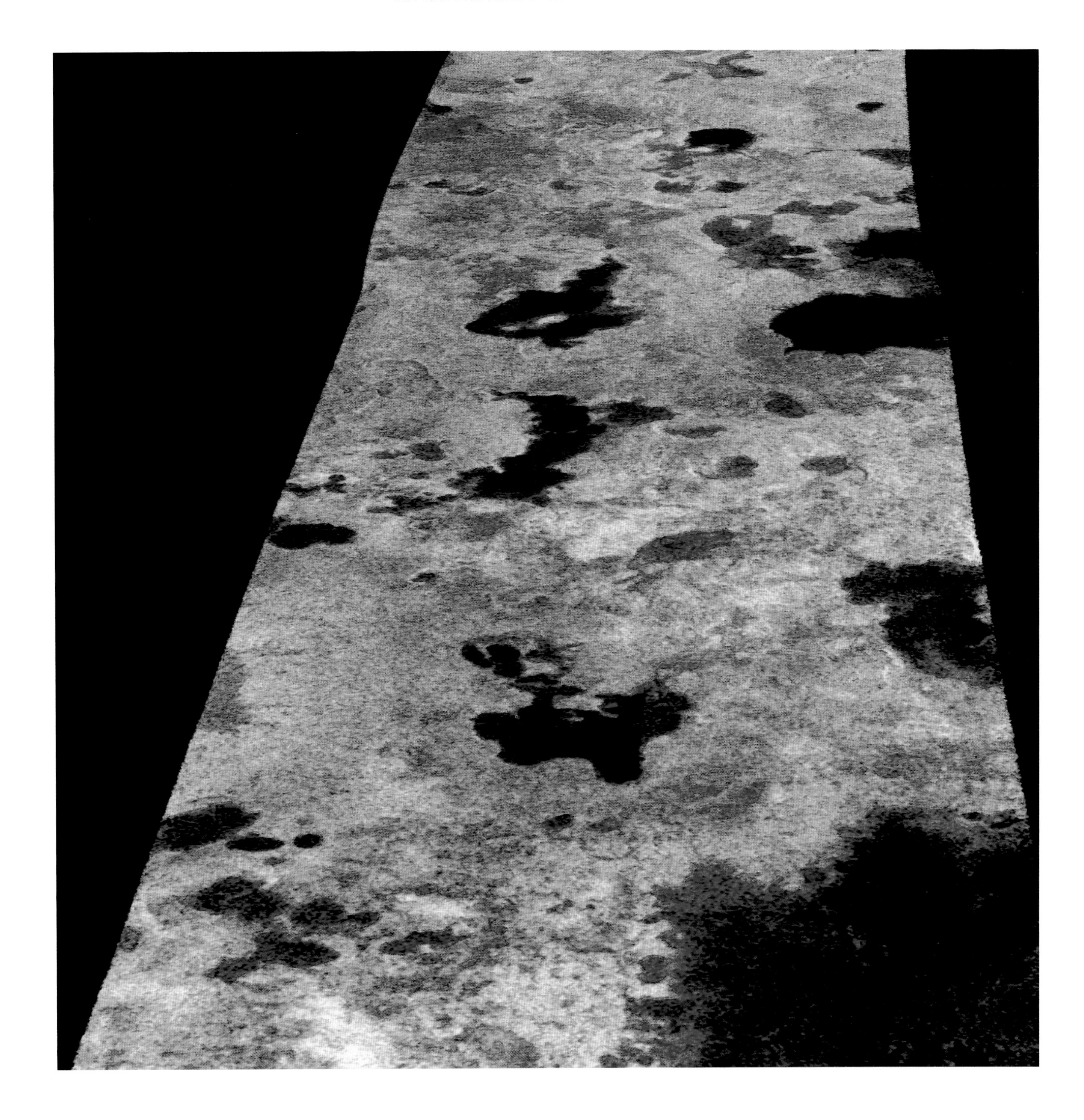

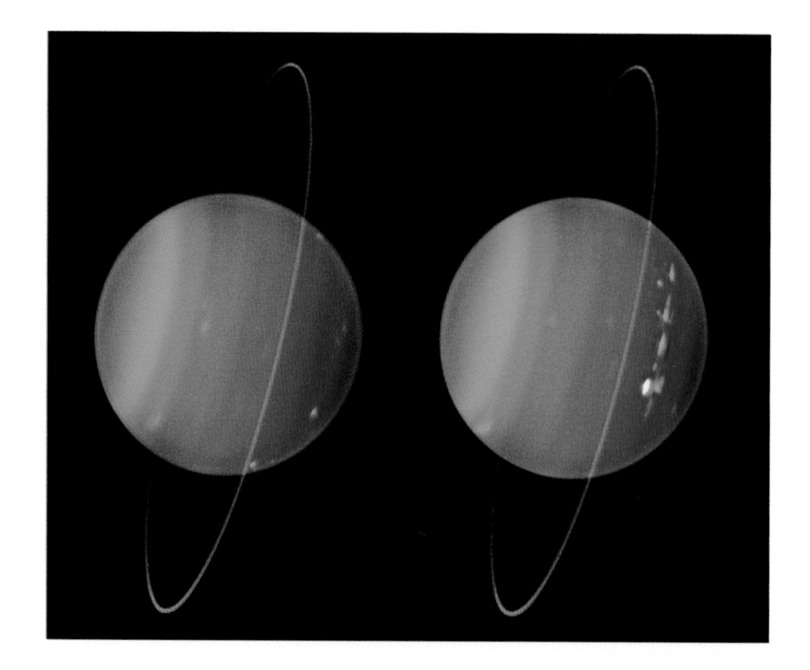

Their cloud top temperatures are much the same, –353°F (–214°C) for Uranus and –360°F (–220°C) for Neptune even though Neptune is so much further from the sun. This is because Uranus, unlike the other three giants, has no internal heat source, or at best a very weak one.

Uranus differs from other giants in another way; relative to the plane of its orbit, the rotational axis is tilted by 98 degrees—more than a right angle. In fact Uranus "rolls" along in its orbit. Sometimes one of the poles is face on to the earth, so appearing to lie in the middle of the disc; at other times it is the equator which is presented. It has long been believed that at an early stage in its evolution the planet was hit by a massive body, and literally tipped over, but this does not sound very plausible, particularly in view of the fact that the largest satellites move in the plane of the Uranian equator. According to a new theory, there were complicated interactions involving both the ice giants plus the material which now makes up the Kuiper Belt; one result of which was that the axial inclination of Uranus was gradually increased to its present value. All this leads to a very unusual calendar. Over the course of one Uranian year, each pole has a night lasting for 21 Earth years, and a day of equal length. There is, moreover, another complication. *Voyager 2* made only a single pass of the planet, on January 24, 1984, so that one pole was in sunlight and the other in darkness. The International Astronomical Union rules that all poles above the plane of the ecliptic are north poles, and all those below the ecliptic are south poles. In this case it was the South Pole which *Voyager 2* saw as sunlit. However, the NASA members reversed this, and referred to the sunlit pole as the North Pole. Take your pick!

Voyager passed about 50,000 miles (80,000 km) above the cloud tops, and various cloud patterns were seen, but nothing at all striking. Since then clouds have been imaged by the Hubble Space Telescope, and it has been found that the winds there are strong. Hubble has also recorded Uranus' system of thin, dark rings.

Five satellites were known before the *Voyager* mission; Miranda, Ariel, Umbriel, Titania, and Oberon. All are below 100 miles (160 km) across, and all have icy surfaces, but Miranda has several distinct types of terrain which are not easy to interpret geologically. *Voyager* flew past at only 1,800 miles (2,900 km) above the surface and provided razor-sharp images. Miranda is unlike any other satellite, with cliffs towering to well

Top: These two false-color images taken with an infra-red camera show opposite sides of Uranus, with the planet's "north" pole at four o'clock. The rings can be clearly seen, as can the methane clouds below.

Above: Computer-assembled mosaic of Miranda (*Voyager 2*), January 24, 1986. The surface is remarkably varied. The missing parts of the surface were included in later pictures, which also showed much more detail. Miranda is unlike any other world in the solar system.

Opposite: Neptune, from *Voyager 2*, at a range of 10,000,000 miles (16,093,440 km). The Great Dark Spot is prominent; below is the small bright cloud nicknamed the Scooter. When the Hubble Space Telescope obtained pictures, several years later, the Great Dark Spot had disappeared. Neptune is a dynamic world—much more active than Uranus.

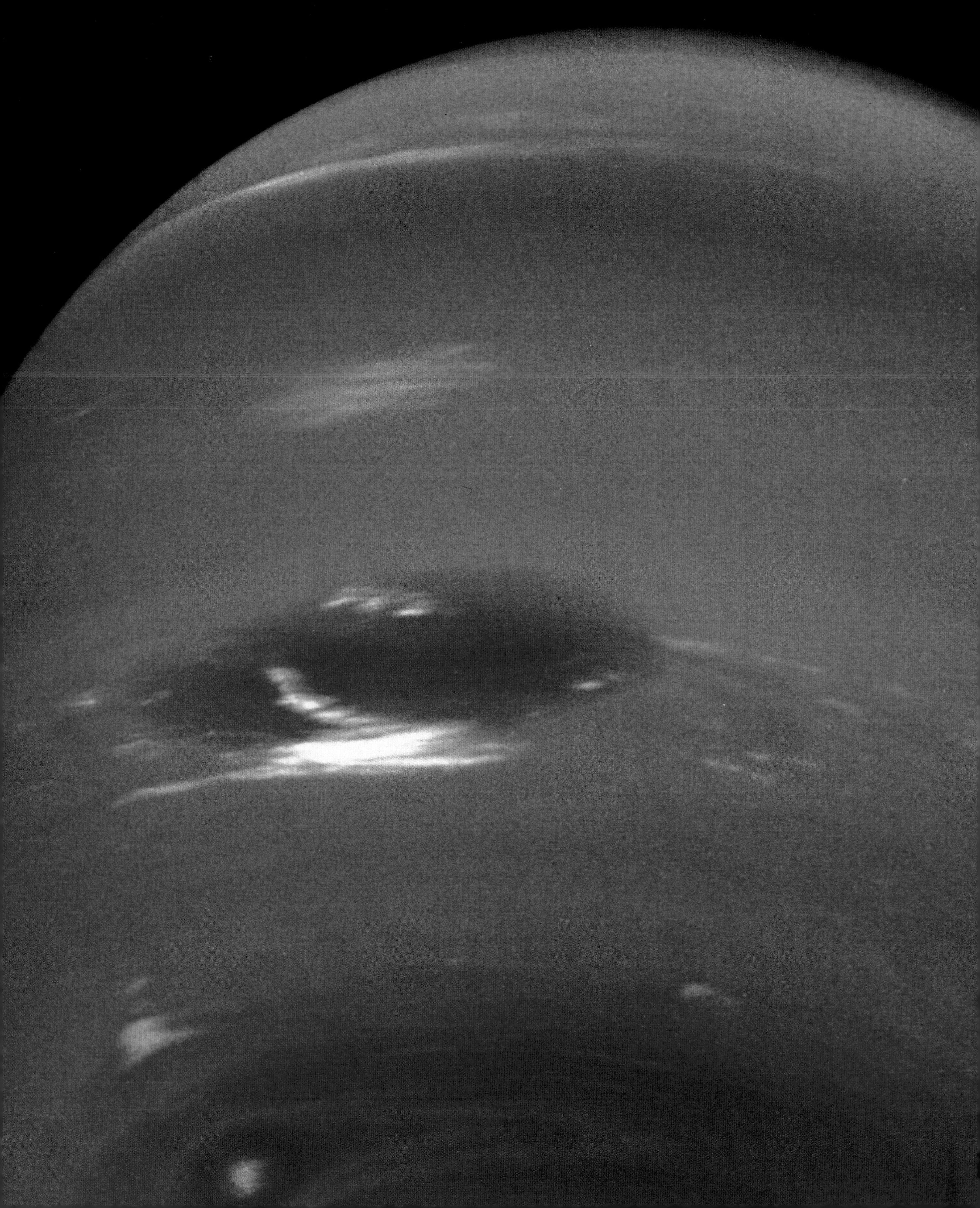

Opposite: Triton, from *Voyager 2* on August 24, 1989, from 330,000 miles (531,084 km). The resolution is 6.2 miles (10 km). Note the dark streaks produced by the nitrogen geysers, and the difference between the pink polar region and the rest of the satellite. This image was made from pictures taken through green, violet, and ultra-violet filters.

over 12 miles (19 km) and large trapezoidal-shaped features, two of which have been nicknamed "the Chevron" and "the Race-Track."

Five years after leaving Uranus, *Voyager 2* encountered Neptune, passing by at a range of just over 3,000 miles (5,000 km). Neptune does not share Uranus' axial inclination, and is a much more dynamic world, with prominent cloud patterns. There was also a huge oval, about 6,000 miles (10,000 km), which became known as the Great Dark Spot, long; its rotation period was over 18 hours, so relative to the adjacent clouds it drifted steadily westward in longitude. Hanging above it were bright cirrus-type clouds made of methane ice, and there was another feature further south, which had a shorter rotation period and was nicknamed "the Scooter." Unfortunately these markings could not be followed once *Voyager 2* had gone on its way, because no earth-based telescope could show them, and the Hubble Space Telescope had not yet been launched.

When Hubble did come into operation, it was found that the Great Dark Spot had vanished, although other details could be made out. Neptune's surface must therefore change relatively quickly. Wind speeds are extremely strong—in fact Neptune is the windiest planet in the solar system.

Hubble confirmed the existence of a magnetic field, and showed thin, very obscure rings, but the pièce de resistance was Triton, the large satellite. It is 1,660 miles (2,670 km) in diameter, larger than Pluto, and moves in a retrograde direction around Neptune. There is a tenuous atmosphere made up chiefly of nitrogen. Predictably, the surface is ice-coated. The south polar region was covered with pink nitrogen geysers capable of sending material up to heights of thousands of feet. Activity of this sort had certainly not been expected.

Everything indicates that Triton is a captured body rather than an original satellite of Neptune. It must surely have been a member of the Kuiper Belt, which strayed close to Neptune and was unable to break free—it may well have been a "binary KBO," as Pluto is now, so that its companion did escape and is now moving in an independent orbit.

We cannot pretend to have a full knowledge of these remote reaches of the solar system. We have learned a great deal, but for the moment we must be content to wait for the launching of new missions. For this research, as in many other branches of astronomy, space methods are of paramount importance, and it is strange to reflect that half a century ago we were only just starting to send rockets beyond the atmosphere.

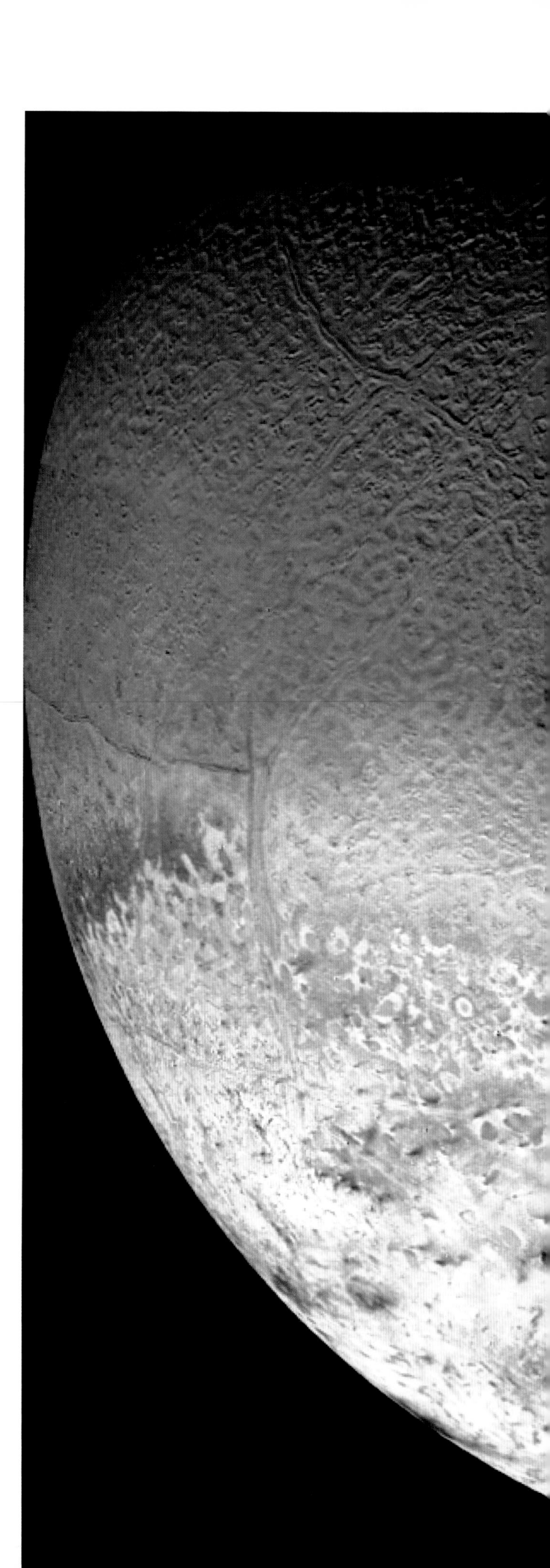

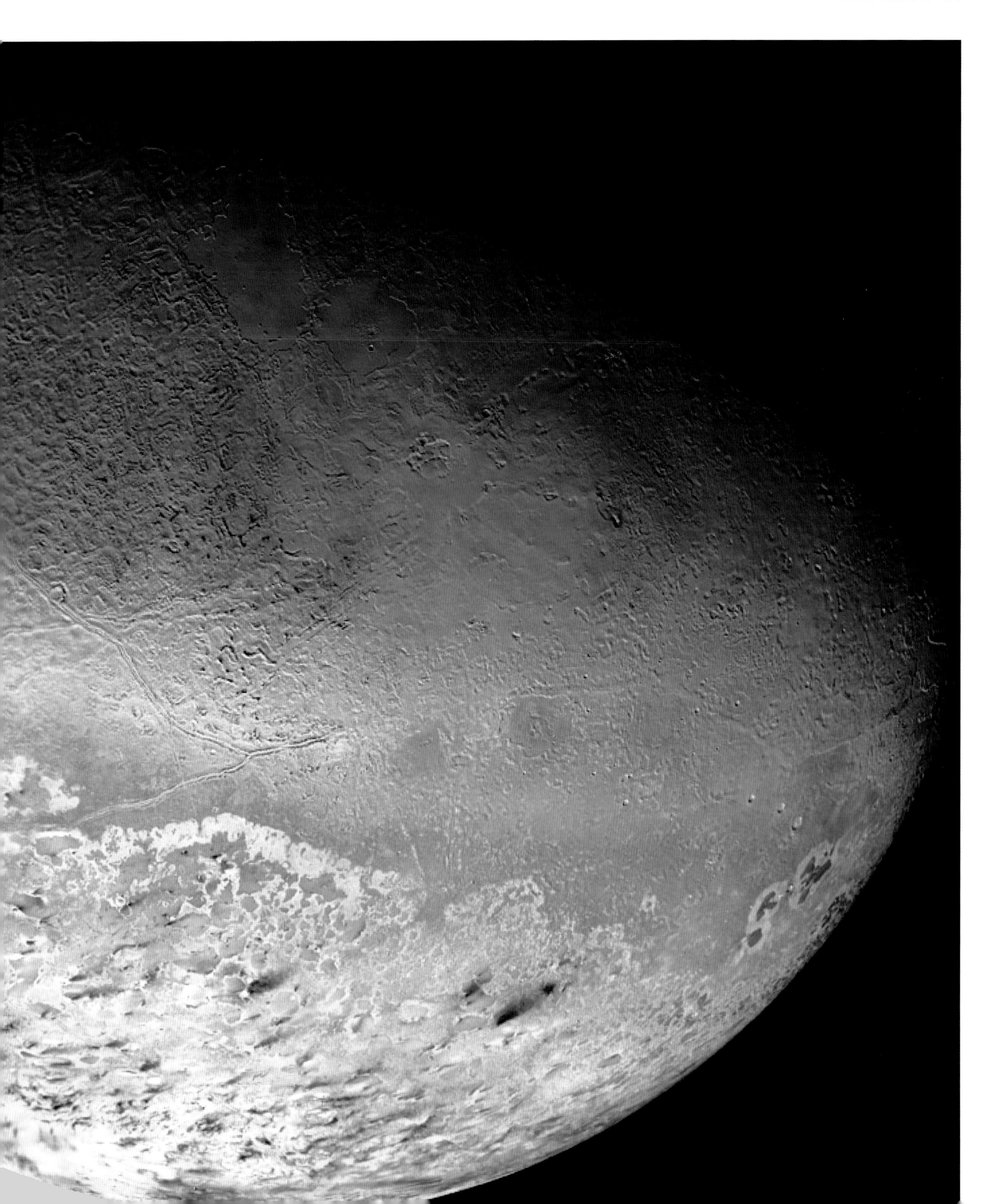

Top left: **1990**
February 11 – Nelson Mandela, prominent anti-apartheid activist and leader of the African National Congress (ANC), is released from prison following a detainment of 27 years. Mandela was sentenced to life imprisonment in 1964 for plotting to overthrow the government.

Top right: **1990**
August 2 – Iraq invades the Gulf state of Kuwait, setting off a chain of events that leads to The Gulf War, a conflict between Iraq and a coalition force of the United States and its allies in order to liberate Kuwait.

Bottom left: **1992**
The Bosnian War claims 100,000 lives between March 1992 and November 1995. Many Bosnian Muslims in Bosnia and Herzegovina are killed in the war under the Bosnian Serb campaign of forced expulsion. Serbs in the Krajina region of Croatia, ethnic Albanians, and later Serbs in the Serbian province of Kosovo are also victims of ethnic cleansing in other conflicts that erupt upon the disintegration of former Yugoslavia.

Bottom right: **1994**
The Channel Tunnel is opened on May 6 by Queen Elizabeth II and François Mitterand. At 31 miles (51.5 km) the rail tunnel links Folkestone, Kent, in England to Coquelles near Calais in northern France.

1990–1999

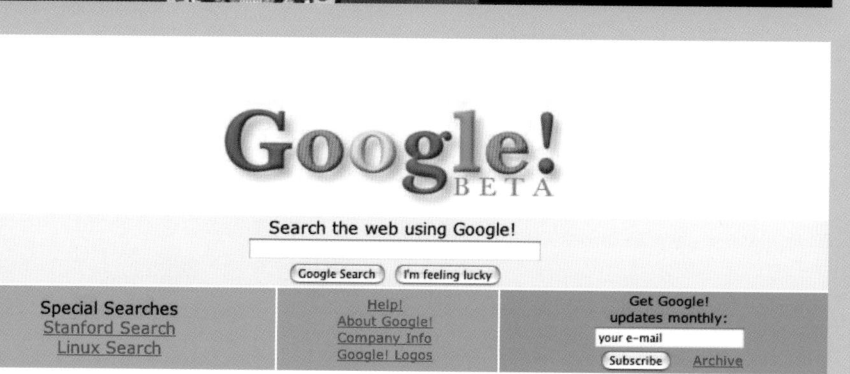

Top left: **1994**
In Rwanda, between April and June 1994, an estimated 800,000 Tutsis and moderate Hutus are killed by Hutu militias after Rwandan President Juvenal Habyarimana, a member of the Hutu tribe, was killed when the plane he was traveling in was shot down on April 6.

Top right: **1996**
July 5 – Dolly the sheep, the first mammal to have been successfully cloned, is born at the Roslin Institute in Midlothian, Scotland. Her birth is eventually announced seven months later and is heralded as one of the most significant scientific breakthroughs of the decade.

Center left: **1998**
News of an extra-marital affair between U.S. President Bill Clinton and 22-year-old White House intern Monica Lewinsky results in the impeachment hearing of the President by the U.S. House of Representatives and his acquittal for charges of perjury and obstruction of justice in a 31-day Senate trial.

Bottom left: **1998**
Google is launched from its first office in a garage in California. The Internet search engine began as a research project by two postgraduate students, Larry Page and Sergey Brin.

Opposite: Comet Wild 2, a composite image obtained by the *Stardust* flyby probe on January 2, 2004. The nucleus of the comet is 3.4 miles (5.5 km) in diameter. The surface is very acrive, jetting gas and dust streams and leaving a tail millions of kilometres long. The orbital period is 6.3 years. *Stardust* collected samples from the comet, and returned them to Earth—the first time that this had been achieved.

09 WANDERERS

MYSTERIOUS VISITORS

Comets are the most erratic members of the solar system. A great comet, with a gleaming head and a tail stretching across the sky, may look magnificent—and yet comets are very insubstantial. I once described a comet as being "the nearest approach to nothing that can still be anything."

A comet is basically a "dirty snowball." The nucleus is made up of rock, dust, and ices; most are less than 20 miles (32km) across. When a comet nears the sun, and is warmed, the ices begin to evaporate, and the comet may develop a tail or tails. Generally speaking the tails are of two main types, both of which point away from the sun. Gas or ion tails are driven out by the solar wind, while dust tails are repelled by the pressure of sunlight. When a comet is moving outward, after perihelion, it travels tail-first. When the comet has receded, the tails shrink and disappear. Many comets do not develop tails of either kind, and although comets are plentiful comparatively few of them reach a point where they are visible to the naked eye.

Comets are true members of the sun's family, but their orbits are different from those of the planets, and are usually very eccentric. Some comets have periods of a few years, so we know when and where to expect them; Encke's comet has a period of only 3.3 years—but since a comet loses mass at every return to perihelion, it wastes away and is short-lived by cosmical standards. We have direct proof of this, because several periodical comets have disintegrated, and have even been seen in the process of breaking up. Only one periodical comet is conspicuous; this is Halley's, which has a period of 75–76 years and last returned in 1986. It will be back once more in 2061, but at the moment it is too faint to be seen even with our most powerful telescopes. Periodical comets of this kind are thought to come from the Kuiper Belt region, but really brilliant comets originate in the Oort Cloud, a collection of cometary objects much further away, at a distance of around 1 light-year. We cannot see the members of the Oort Cloud, which is named after the Dutch astronomer Jan Oort.

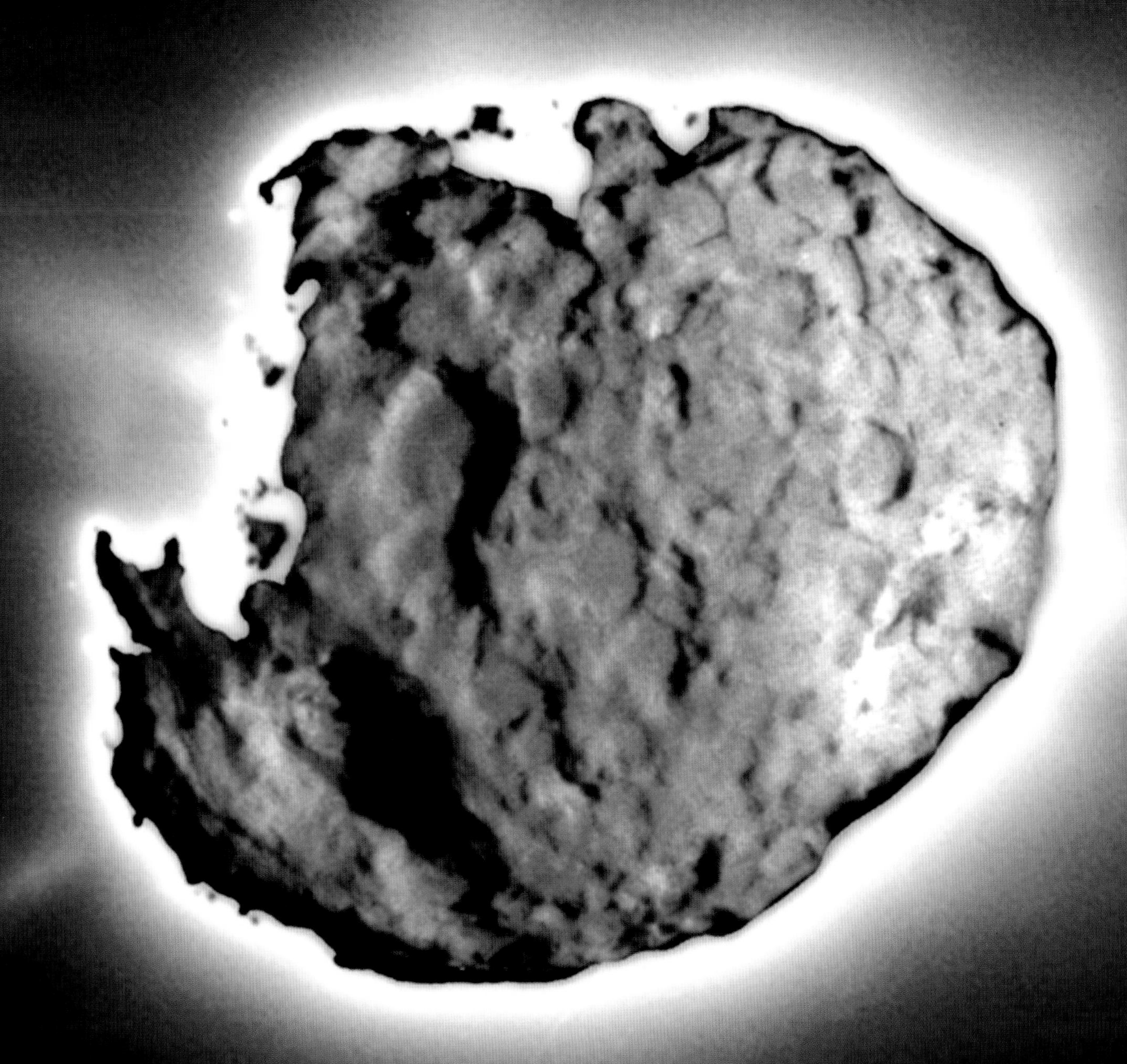

If one of the objects is perturbed for some reason, it starts to swing inwards, finally reaching the inner solar system, where we can see it. One of several things may then happen. The comet may simply swing around the sun and travel back to the Oort Cloud, not to be seen again for many centuries. It may fall into the sun, and be destroyed. Or it may be captured by the pull of a planet (usually Jupiter) and be forced into a short-period orbit, as Halley's has been.

Comets have been recorded since early times, and used to cause great alarm; indeed, the fear of comets is not dead even today. Certainly a great comet looks imposing, and some (such as the comet of 1882) cast strong shadows, but the twentieth century did not produce many of these startling visions. However, there were two notable comets as the century neared its end. The first was discovered in January 1996 by the Japanese amateur astronomer Yuji Hyakutake; it became visible to the naked eye in March, but did not stay bright for very long. It was at its best at the end of the month, with a nucleus of magnitude 0 and a 35-degree tail and was particularly beautiful because of its green color. Actually it had a small nucleus, but came within 10,000,000 miles (16,093,400 km) of Earth, which ranks as a "close approach." It has long since been lost to view; remember to look out for it when it next returns, in about 72,000 years' time!

Lovely though it was, Hyakutake was outmatched by the Great Comet of 1997, discovered independently by two American observers, Alan Hale and Thomas Bopp. This was a giant by cometary standards, with a 40km (25 mile) nucleus. It developed a curved dust tail and a straight ion tail; it became as bright as any star apart from Sirius, and it

Above: Comet Hale-Bopp. It was large by cometary standards, with a nucleus 25 miles (40 km) across; there were both plasma and dust tails plus a tail made up of sodium. It was visible to the naked-eye for over a year. Perihelion was passed on April 1, 1997; unfortunately the comet did not approach Earth closely. It will be back in 2360 years.

Opposite left: Portion of the Bayeux tapestry, showing Halley's comet before the Battle of Hastings (1066). The legend reads, "These men wonder at the star." Evidently the Saxons regarded the comet as a bad omen; King Harold is toppling on his throne! The comet passed perihelion on March 20, 1066, and apparently the nucleus was as bright as Venus.

Opposite right: Wide-angle photograph of Halley's comet in 1910; the bright object to the upper right is Venus. Perihelion fell on April 20, 1910; the closest approach to Earth was on May 20 , at a range of 13,000,000 miles (20,921,472 km). The tail was 140 degrees long. The comet was impressive, but could not match the Daylight comet of the preceding January.

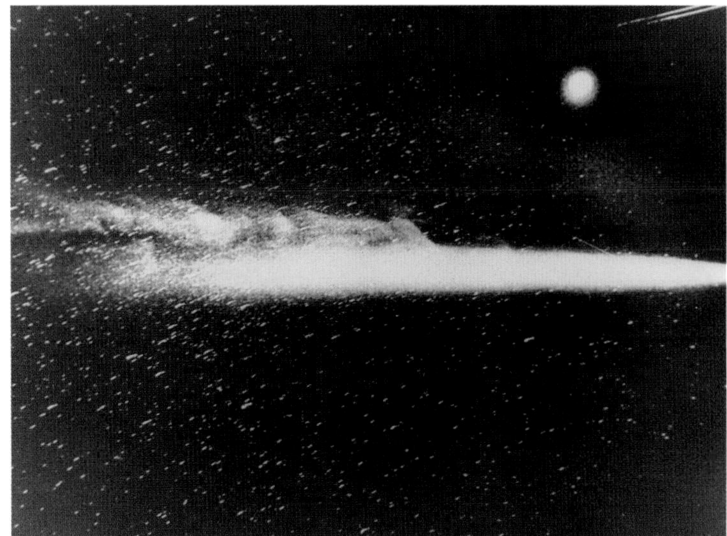

hung in the sky for over a year, remaining visible to the naked eye from July 1996 to October 1997, which is easily a record. It must have been the most photographed comet in history. Unfortunately it did not come within 40,000,000 miles (64,373,760 km) of the earth and we will not see it again, as the period has been computed to be 2,540 years.

COMETARY RENDEZVOUS

The most famous of all comets is named after Edmond Halley, Britain's second Astronomer Royal, who saw it in the year 1682. At that time it was not realized that comets move around the sun, but Halley found that this new comet moved in the same way as comets seen in 1607 and 1531. He predicted that it would be back in 1758, and it duly appeared, reaching perihelion in 1759. Since then it has returned in 1835, 1910, and 1986. At its last return five spacecraft were sent to it: two Japanese, two Russian and one European (the Americans withdrew for financial reasons). Europe's probe, named *Giotto* after the famous artist, penetrated the comet and sent back pictures until close to the nucleus, when the camera was damaged by the impact of a particle probably about the size of a grain of rice. The nucleus was shaped rather like a peanut, 18 miles (29 km) long, and coated with dark material, with jets coming from below. *Giotto* survived the encounter and went on to rendezvous with another comet, Grigg-Skjellerup, although it sent back no more pictures.

Since then there have been several missions to comets. *Stardust*, launched from Cape Canaveral on February 7, 1999, flew past a small periodical comet, Wild 2, on January 2, 2004, sending back pictures from a range of 149 miles (240 km). It did more than this; as it flew past, moving at a relative speed of 14,000 mph (22,000 km/h), it collected samples of the cometary material by catching them in an aerogel-filled shield shaped rather like a tennis racket. (Aerogel is a silicon-based solid with a porous, spongelike structure; it is a thousand times less dense than glass, so on impact the particles could burrow inside it.) The samples were stored in a capsule; *Stardust* returned to the neighborhood of the earth, and the capsule was parachuted back to the ground, landing in Utah on January 15, 2006. Analysis of the samples showed the expected amount of ice, but also some minerals of types which form at high temperatures. Since the comet has presumably never been heated, these minerals are decidedly puzzling. The close range images of the comet showed the usual dark, cratered, surface, with jets of evaporating material. *Deep Space 1*, sent up from Canaveral on October 24, 1998, went to another small comet, Borrelly's, which had an orbital period of 6.8 years, and has made the most returns ever since it was discovered in 1904 by the French astronomer Alphonse Borrelly. *Stardust* reached the comet on September 21, 2001, and transmitted images of a most extraordinary object, 8 miles (13 km) long, with an exceptionally dark coating; there were smooth areas, jumbled hills, and the usual spurting jets.

NASA began a new experiment on January 12, 2005, when the *Deep Impact* probe was launched. The target was yet another small, well-known comet, Tempel, 1 which moved round the sun in a period of 5.5 years and had been first seen by Ernst Tempel as long ago as 1867. The plan was to release a copper impactor, about the size of a household refrigerator, and crash it into the comet, while the spacecraft itself veered prudently out of the way and prepared to take pictures. All went well. On July 3 the probe landed—not gently, but at tremendous speed—and produced a 300 ft. (90 m) crater; needless to say, the

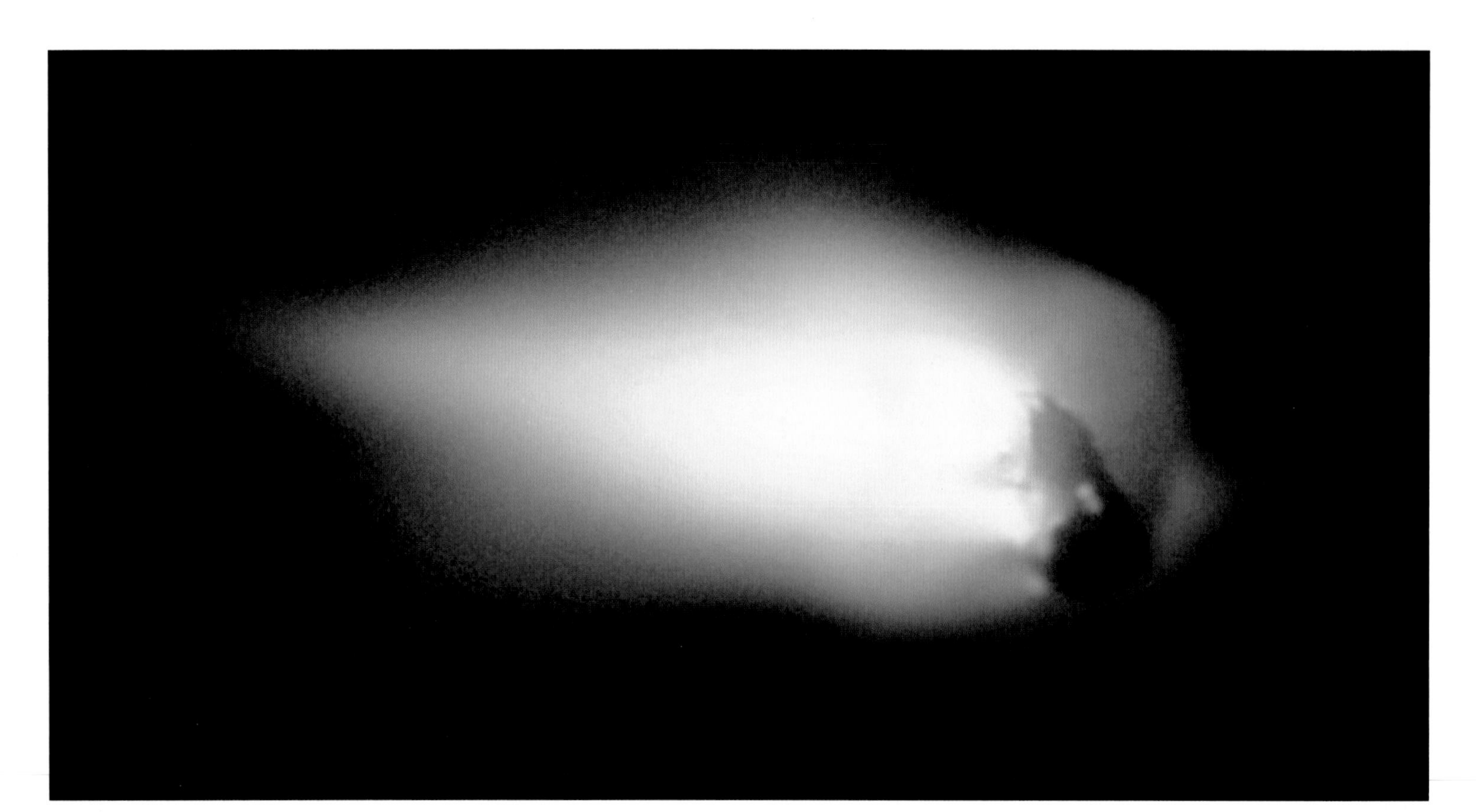

copper missile was instantly vaporized. The surface of the comet was shown to be covered with dust, as fine as talcum powder, while the nucleus of the comet had a fluffy structure, weaker than a bank of unconsolidated snow. The gravitational pull is so weak that if an astronaut could land there, he would be able to jump free of the comet altogether. Incidentally, there were some people who objected to the whole idea of hitting a comet, for fear of damaging it. In fact, the effect was about the same as would be caused by throwing a baked bean at a charging hippopotamus. Astrologers were particularly vocal, since destroying a solar system body might upset their horoscopes!

There is a good chance that some of the tiny close-approach asteroids may be the nuclei of comets that have lost all their volatiles, and we have seen comets in their death-throes. Biela's Comet, which had a period of 6.8 years, broke in half during the return of 1846; the pair returned on schedule in 1852, but that was their last appearance, although their remnants manifested themselves in the form of meteors. We have had another case of disintegration much more recently; Comet Schwassmann-Wachmann 3 is breaking up and its life expectancy is very limited indeed.

One great comet—that of 1882—was responsible for a whole new branch of astronomical research. Photography in those days was comparatively rudimentary; reasonable pictures of the moon had been obtained, but comets presented much greater problems. There were no images of the earlier nineteenth-century visitors, which was a great pity. The two brightest were those of 1811 and 1843, both of which were easily visible with the naked eye in broad daylight. The comet of 1811, discovered by Honoré Flaugergues, had a core over 1 million miles (1.6 million kilometers) in diameter—much larger than the sun—and a straight tail which stretched for over 16 million miles (25 million kilometers). We do not know the size of the nucleus. The comet of 1843 was even brighter, but for sheer beauty it was surpassed by Donati's comet of 1858, which had a wonderfully curved main tail and two smaller ones. The orbital period may be around 2,000 years. One low-quality photograph of it was taken by an English commercial photographer, William Usherwood, with a f/2/4 focal ration portrait lens, but when the Great Comet of 1882 became really spectacular David Gill, Director of the Cape Observatory in South Africa, set out to obtain a much better picture. He succeeded—and also recorded large numbers of stars. Instantly he realized that the best way to map the sky was to use photography, and this led on to the first photographic star atlases.

It is a pity that we have had no comparable comets recently. The last was in 1910, and was nicknamed the Daylight Comet; it appeared a few weeks before Halley's, and was much brighter. Hyajutake, Hale-Bopp and some others have been conspicuous, but not brilliant enough to cast shadows. Yet a new Great Comet may appear at any time, and if we have enough warning attempts will certainly be made to send probes to it.

A collision between the earth and a comet would have catastrophic results—we have only to look at what happened when Shoemaker-Levy 9 hit Jupiter in 1996. Of course, this refers only to a cometary nucleus; the tails are much too rarefied to cause any damage at all, even though they contain gases which would be deadly at higher concentrations. The tails are millions of times less dense than the air we breathe; in 1861 the earth passed through the tail of Tabbott's comet, and skimmed the tail of Halley's comet in 1910, without any detectable effects. The object which landed in Siberia in 1908 may well have been the nucleus of a small comet. But meteorites are also to be taken into account.

METEOROIDS AND METEORITES

Quite apart from planets, comets, and natural satellites, the solar system is far from empty. There is, of course, a great deal of "dust," but there are also bodies of appreciable size that may land on Earth and produce craters. Once they have reached the earth's surface they are known as meteorites; most museums have collections of them. Some are immense; the Hoba West meteorite in Namibia is still lying where it fell, in prehistoric times, and moving it would be difficult, since it weighs over 60 tons. (This is a troubled area, and United Nations peacekeepers were detailed to safeguard the meteorite—only to be removed when they were shown to be vandalizing it.) One crater, on the uninhabited Devon Island in the Canadian Arctic, has been used as a temporary research base by NASA scientists, because it seems to be the closest terrestrial equivalent to conditions on Mars. Haughton Crater is 12 miles (19 km) across, and is thought to have been formed by a meteorite 23,000,000 years ago, at the end of the geological period known as the Oligocene. This was long before the appearance of the first primates, but modern-type mammals did exist.

The impactor was probably about 1.5 miles (2.5 km) in diameter—this is larger than some of the near-Earth asteroids which have been observed. Both come from the main asteroid belt, and it is important to note that there is absolutely no connection between meteoroids and shooting-star meteors, which, as we have seen, are of sand-grain size and are nothing more than cometary debris.

The best-known impact crater, in Arizona, about 35 miles (55 km) east of the town of Flagstaff, is 4,000 ft. (1,200 m) across and 570 ft. (175 m) deep; it is always called Meteor Crater, though it really should be Meteorite Crater. Its age is now believed to be 50,000 years—it is believed to have struck during the Pleistocene period, when the climate on the Colorado Plateau was much cooler and damper than it is now; the area was open grassland, and the descent of a meteorite at least 150 ft. (45 m) across must have dismayed the local inhabitants such as woolly mammoths, giant ground sloths and the like. It is estimated that the explosion dug out 175 million tons of rock, and all life within a radius of several kilometres would have died instantly. Meteor Crater is a noted tourist attraction, and is easy to reach, not far off Highway 99.

Could this sort of impact happen again? The answer must be "yes." The Siberian impactor of 1908—whether asteroid, cometary nucleus, or cometary

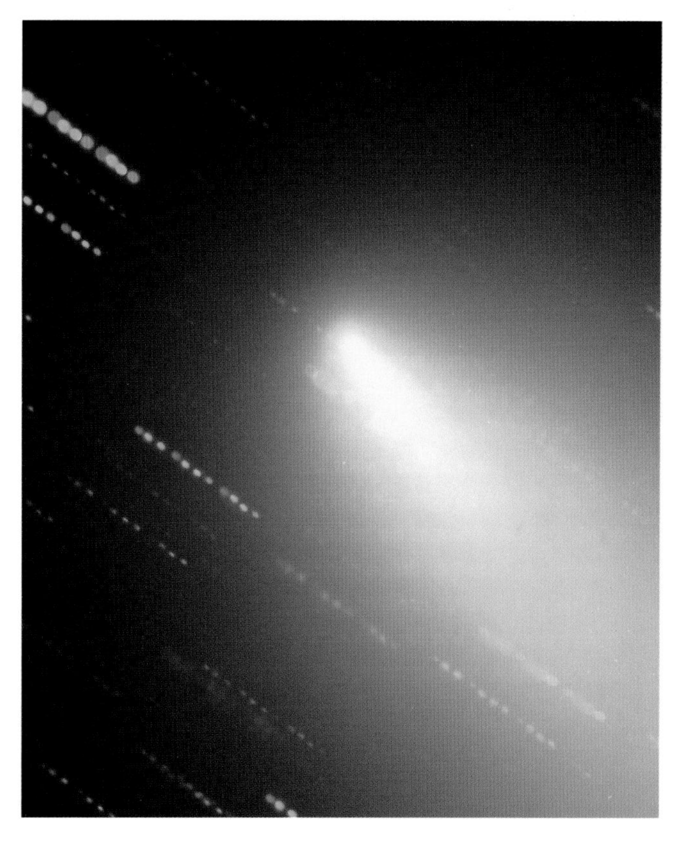

Opposite: Nucleus of Halley's comet, imaged by the *Giotto* spacecraft on March 14, 1986; the probe had penetrated the comet's coma, and passed within 380 miles (612 km) of the nucleus, which was shaped like a peanut, with a longest diameter of 9.3 miles (15 km). The nucleus was dark-coated, with active dust-jets. *Giotto* survived the encounter.

Top: The Great Comet of 1882, photographed by David Gill from the Cape in South Africa. This was the first good photograph of a comet. It also showed many stars. The comet was visible to the naked eye in May and June 1882, with a strong yellowish color. It has an orbital period of many centuries.

Above: A disintegrating comet—73P/Schwassmann-Wachmann 3, which had an orbital period of 5.4 years. In 1993 it began to break up. This picture, taken with the Very Large Telescope in Chile (four filters) shows broken pieces surrounding one of the larger fragments. As the telescope was tracking the comet, the stars appear as colored trails.

fragment blew pine-trees flat over a wide area, and it was sheer luck that it landed in an uninhabited tundra. (There is no authenticated case of anyone being killed by a tumbling meteorite, although admittedly several people have had narrow escapes.) On the other hand, it is widely believed that "extinction," when many Earth-creatures died at the same time, was caused by cosmical bombardment. Around 65,000,000 years ago, the dinosaurs, which had been dominant for so long, disappeared abruptly, and this has been linked with the fall of a 6 mile (10 km) meteorite leaving a now buried crater at Chicxulub, in the Yucatán Peninsula in Mexico. Much earlier, around 250,000,000 years ago—at the end of the geological Permian Period—there was an even greater extinction, when over 90 percent of animal life on Earth vanished. Space research has played a role in investigating this disaster. Using results from a NASA satellite designed to measure gravitational data, Ohio State University scientists have identified a 300 mile (500 km) crater in Wilke's Land, East Antarctica, hidden more than a kilometre below the ice sheet. If the Ohio team members are right—and there is no reason to believe otherwise—the impactor was 30 miles (50 km) wide and its effects would have been felt over the entire planet.

If we saw an asteroid of this size heading towards us, on a collision course, there is frankly nothing that we could do about it. Fortunately, the chances of a really major impact in the foreseeable future are very slight indeed.

When Yuri Gagarin became the first man to fly above the atmosphere in 1961, there were many unknown factors. Pessimists suggested that Gagarin might be killed by cosmic rays or harmful emissions from the sun; he might be hopelessly spacesick; he might be blinded by the harsh light—or his craft might be battered by wandering particles, so that it would be holed and the capsule would be depressurized. None of these things happened, but on journeys to other worlds there is an ever-present danger of an impact. In the future, when travel to the moon and Mars may well become commonplace, we must reluctantly realize that sooner or later there will be a tragedy of this sort.

When the *Pioneers* and then the *Voyagers* first passed through the asteroid belt on their way to the outer planets, there were uneasy feelings at NASA. All the large members of the Main Belt could be avoided, because their orbits were well known, but at that time there was no way of finding out how many tiny bodies existed, and a spacecraft could be fatally damaged in a collision with an object the size of, say, a teapot. Up to now this has not happened. Either we have been exceptionally lucky or else, more probably, the danger from large meteorites is less than had been feared.

We can keep clear of the chief meteor streams, because we know their parent comets—notably the August Perseids (Comet Swift-Tuttle), the November Leonids (Temple Tuttle), the December Ursids (Tuttle), the autumn Taurids (Encke), and the October Draconids (Giacobini-Zinner). There are two streams associated with Halley's comet—the Eta Aquarids of May and the Orionids of October—while the December Geminids are linked with the tiny asteroid Phaethan, which may well be an ex-comet. (It is notable that three parent comets bear the name of Tuttle. Horace Tuttle was a nineteenth-century astronomer and Naval officer who carried out a great deal of useful work and fought with bravery during the American War of Independence.)

What are termed micrometeorites—bodies too small even to cause luminous effects when they dash into the atmosphere—pelter spacecraft all the time, and "dust," lying in or near the main plane of the solar system, is responsible for the lovely glow of the Zodiacal Light. But at least the prophets of doom, who were convinced that travel above the protective atmosphere would never be possible, have been shown to be wrong.

We have reached the moon, our robot craft have made controlled landings on Mars and Venus; we have penetrated the heart of a comet, touched down upon an asteroid, and sent our rockets to survey all the planets from close range. We cannot travel to the planets of other stars, but space research is telling us what we might expect to find. In a different way, we are exploring our galaxy—and others.

Above: Meteor Crater, Arizona (it should really be named Meteorite Crater). It is around 50,000 years old and almost 4,000 ft. (1,219 m) in diameter, with a depth of almost 600 ft. (183 m). It is a well-known tourist attraction, and is very easily reached, not far off Highway 99. It is the best-preserved of all large impact craters.

Opposite: "Diamond Ring" effect at the end of totality. This picture was at the total eclipse of July 11, 1991, best seen from central America or Hawaii. As the first tiny portion of the sun reappears, there is the lovely ring effect shown here. It is very brief; then, as the moon moves away, the light floods back over the landscape, the sky brightens, the corona disappears, and the glory of totality is over.

10
THE REALM OF THE STARS

Up to now we have been discussing the members of the solar system whose distances are measured in millions of miles. When we come to consider the stars, the situation is very different, because even the nearest star beyond our system is over four light-years away—that is to say, around 24 million million miles (39 million million kilometers). A star looks like a mere point of light. Remember the old rhyme:

Twinkle, twinkle, little star,
How I wonder what you are.

There is no mystery about the twinkling; it is due to the earth's unsteady atmosphere. But to investigate bodies that are so far away is indeed a daunting idea.

THE SUN

However, there is one star which is close enough to be examined in detail, and that is the sun. It is a mere 93 million miles (150 million kilometers) from us, and even a small telescope will show the dark patches known as sunspots, produced when the sun's lines of magnetic force break through to the surface and locally cool it. (A word of warning here. Never look straight at the sun through any telescope or pair of binoculars, even with a dark suncap fitted over the eye piece; permanent blindness may result, and, sadly, this has happened quite a number of times in the past. There are ways around this, such as the method of projection, but there is one golden rule about looking straight at the sun through a telescope: don't.)

The glorious corona, which surrounds the sun is made up of very rarefied gas, and is visible to the naked eye only during a total solar eclipse, when the moon passes in front of the sun and hide it briefly. It is sheer luck that the two bodies look the same size in the sky; the sun's diameter is 400 times that of the moon, but the sun is al

400 times farther away. To find out more about the sun, we must turn to the spectroscope. What we call "white" light is not really white at all; it is a mixture of all the colours of the rainbow. It was Isaac Newton who first passed a beam of sunlight through a hole in a screen, then passed it through a glass prism and produced the first solar spectrum, with red at the long-wave end and violet at the short-wave end, with orange, yellow, blue, and green in between. Light is a wave motion and the color depends upon the wavelength.

Over a century later, the German optician Josef Fraunhofer produced a more detailed spectrum, and found that the rainbow beam was crossed by dark lines which never varied either in position or in intensity (they are still often called "Fraunhofer Lines"). Fraunhofer did not know what caused them; they were eventually explained by two of his countrymen, Gustav Kirchhoff and Hubert Bunsen.

An incandescent solid, liquid, or gas at high pressure will yield a rainbow or continuous spectrum, but a gas at low pressure will produce isolated bright lines, each of which is the trademark of one particular element or group of elements. For example, two prominent yellow lines in a particular position must be due to sodium, nothing else. The sun's bright surface or photosphere yields a continuous spectrum. Surrounding it is the low pressure solar atmosphere, which yields a line or emission spectrum. Seen against the background rainbow, these emission lines are "reversed," and appear dark. In the yellow part of the band there are two dark lines which correspond to the bright lines of sodium, telling us that there is sodium in the sun. By now most of the 92 naturally occurring elements have been identified there. Of course, the spectrum is complicated; iron alone can produce thousands of lines.

The stars are suns, and show spectra of the same type. But before dealing with them, let us pause to show what space research methods have told us about our own sun.

We must remember that the sun sends out radiation at all wavelengths, not merely in the visible range, and that most of the sun's radiations are blocked by the earth's atmosphere. Also, our only chance of seeing the solar corona really well is during the fleeting moments of a total eclipse, although spectroscopes make it easy to study the prominences at any time. The advantages of space methods was realized at an early age, and in fact the first positive results date back to 1946, when ultraviolet spectra of the sun were obtained from a camera carried to a height of 34 miles (55 km), on one of Wernher von Braun's V2 rockets. It was promising, but a rocket can stay aloft only briefly, and really intensive research had to await the development of artificial satellites.

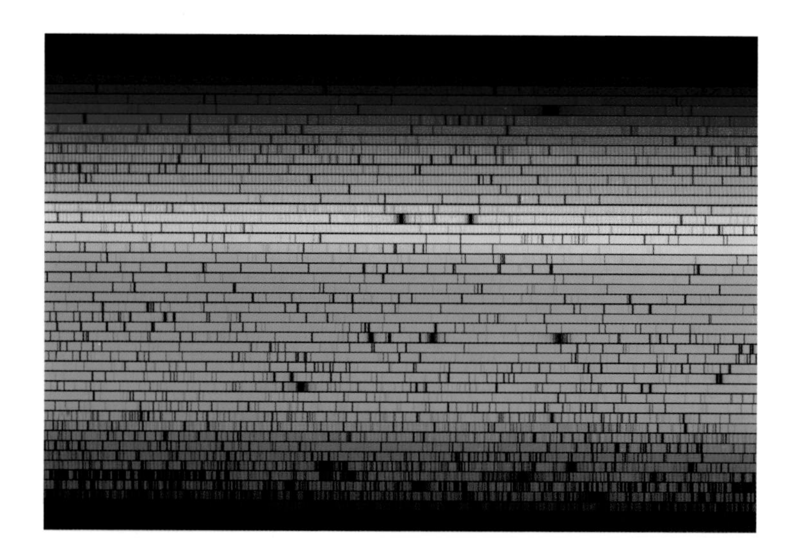

Above: The solar spectrum spectrometer at the Kitt Peak National Observatory, USA. Note the dark "Fraunhofer Lines," which give us vital information about the sun's composition. Each of the dark lines is the trademark of one particular element or group of elements.

THE SUN FROM SPACE

Several OSOs (Orbiting Solar Observatories) were launched from 1962. Astronauts manning the first NASA space station, Skylab, from 1973 to 1974, paid great attention to the sun, and obtained splendid images; loops and prominences were seen to tear loose from the solar surface to form what are now known as coronal mass ejections—which are of fundamental importance in all studies of the sun. Since then many other solar missions have been launched, but four are of special note: *SMM* (Solar Maximum Mission), *Ulysses*, *Yohkoh*, and *SOHO*. (Note that none of these passed really close to the sun; the record approach is still held by a small German probe, *Helios 2*, which ventured to a distance of 27 million miles/43 million kilometers in 1976.)

SMM was launched on February 14, 1980, and was scheduled to remain operational as the sun reached the peak of its 11-year cycle of activity. It produced valuable data over a wide range of wavelengths,

Above: A Coronal Mass Ejection (CME); a sequence extending over two hours, on March 20, 2000, imaged by instruments on the Solar and Heliospheric Observatory spacecraft. A CME is a huge "bubble" or plasma ejecting thousands of millions of particles into space. The results of a CME include magnetic storms and brilliant displays of aurorae.

and lasted until November 1989, but it had a checkered career; faults developed at an early stage, and in 1984 astronauts on the *Challenger* space shuttle captured it, repaired it, and released it back into orbit. *Ulysses* (1990) was sent to survey the poles of the sun, which are out of view from earth because we always see the sun broad side on, so to speak; *Ulysses* was first to make the long trek out to Jupiter, and use the gravitational force of the Giant Planet to oust it into the required orbit. The Japanese *Yohkoh* (launched 1990) was concerned mainly with the corona; it obtained spectacular pictures, and helped towards an understanding of the structure and the way in which the corona is heated. Yohkoh lasted for most of a solar cycle, and did not cease transmitting until 2001. But perhaps the most successful spacecraft so far has been SOHO, the Solar and Heliospheric Observatory, which was sent up on December 2, 2005, and is still functioning perfectly at the time of writing (2007). It stays near what is called the Lagrangian pull point, where the sun's pull balances that of the earth; it lies almost 1 million miles (1.6 million kilometers) away from us, and has the same orbital period. It is always in sunlight, and always well placed for communicating with earth. It carries a whole battery of instruments, capable of studying the sun in all its aspects. It has also discovered a surprising number of comets which swing close to the sun, many of which do not survive. In 2005 SOHO found its thousandth comet.

THE GALAXY BEYOND

Without a good knowledge of solar physics, we could not hope to learn much about the stars, but we always have to contend with the earth's atmosphere, and this means using spacecraft. It is not true to say that space astronomy has superceded ground-based astronomy; the two are complementary. However, missions beyond the atmosphere during the past few decades have been immensely significant. A major success came in 1983 with IRAS, the Infrared Astronomical Satellite, a joint project between Britain, the Netherlands, and the United States. IRAS mapped 96 percent of the sky at infrared wavelengths during its 10 month lifetime, and discovered about ½ million discrete resources, but perhaps the most striking "find" was that of a dust disc around the brilliant star Vega, possibly indicating the presence of a system of planets. The International Ultraviolet Explorer (IUE) carried out an equally thorough survey at the short-wave end of the electromagnetic spectrum, and astronomers made well over 100,000 observations of objects ranging from planets, comets, supernovae, and galaxies to remote quasars. IUE began its career on January 26, 1978, and was expected to last for up to three years. In fact it was still operating perfectly in 1996, when it was switched off at the behest of accountants. Europe's Hipparcos (High Precision Parallax Collecting Satellite) was active from 1989 to 1993, and sent back accurate measurements of star positions, which led to the publication of the Millenium Star Atlas in 1997, including more than a million objects.

HUBBLE

Pride of place must surely go to the Hubble Space Telescope (HST), named after the great twentieth-century astronomer Edwin Hubble who first discovered that the galaxies are separate star systems, millions of light-years away, and that the entire universe is expanding.

The Hubble Space Telescope, launched from Cape Canaveral on April 24, 1990, is a reflecting telescope of the type known as a Ritchey-Chrétien; it has a 94 in. (239 cm) mirror with a focal length of 189 ft. (57.6 m). It moves around the earth at a height of 370 miles (600 km), and when passing over is easy to see with the naked eye if you know when and where to look for it. The orbital period is 96 minutes. Seeing conditions are perfect; Hubble can operate at ultra-violet, visible and near infrared wavelengths.

Since the repair of Hubble's faulty mirror in 1993, worlds of all kinds have been imaged. Active volcanoes on Jupiter's satellite Io were seen for the first time since the Voyager passes of over a decade earlier. Beautiful pictures were obtained of Mars and Saturn; shadings could be made out even on Pluto. HST watched the impact of the comet Shoemaker-Levy 9 on Jupiter, and imaged the nucleus of the lovely comet Hyakutake in 1996. But it was beyond the solar system, in our galaxy and beyond, that Hubble really came into its own.

Betelgeuse in Orion is a vast red super giant star, 11,000 times as luminous as the Sun, but at its distance of 430 light-years its apparent diameter is very small. In March 1996, HST took a picture showing genuine detail; there is an extended atmosphere, imaged in ultraviolet light, and on the star's surface there is an enormous bright spot over

Opposite: The Crab Nebula, M.1, in Taurus. This is the remnants of the supernova seen in 1054, which becomes visible with the naked eye in broad daylight; it is 6,000 light-years away, and emits over the whole range of the electromagnetic spectrum. This picture, taken by the Hubble Space Telescope, shows the intricate structures.

ten times the diameter of the earth. The resolution of the telescope is equivalent to seeing a car's headlights separately from a range of 6,000 miles (10,000 km).

Betelgeuse is variable in light, but its fluctuations are mild compared with those of another Hubble target, Eta Carinae, in the southern hemisphere of the sky. It is about 10,000 light-years away, and one of the most massive and most unstable stars known. For a few years in the nineteenth century, it shined more than any other star in the sky apart from Sirius. It then declined, and for many decades now it has hovered on the brink of being visible to the naked eye, but the fading is more apparent than real, because most of the energy is sent out in the infrared, and the total luminosity is over 4,000,000 times that of the sun.

Telescopically, Eta Carinae looks quite unlike a normal star; when I first saw it I remember describing it as "an orange blob." Hubble images show billowing clouds of dust and gas. There have been suggestions that Eta is a close binary made up of two components that are very close together, but on the whole it seems more likely that we are dealing with a single, super massive star which is highly unstable; there is an expanding shell, with a lobe-like structure. The reddish color can be explained by fast-moving nitrogen atoms ejected from the star. There is no doubt that in the future—perhaps tomorrow, perhaps not for a million years—there will be a grand collapse followed by a cataclysmic explosion. Most of the star will be blown away in what we call a supernova outburst, leaving a tiny remnant made up of neutrons. A cupful of neutron star material would weigh as much as an ocean liner!

In 1054 a brilliant supernova was seen in the constellation of Taurus, the Bull. It became visible with the naked eye in broad daylight. There were no telescopes in those days, and when the star had faded below the sixth magnitude it was lost, but today we see its remnant—the Crab Nebula, which is 6,000 light-years away. Good binoculars will show it, but Hubble brings out the intricate details, and shows the tiny neutron star in the middle. This star is rotating rapidly, and sending out pulses of radio waves, which we can detect. Thousands of these "pulsars" are now known, though not very many can be identified without optical objects.

A star such as the sun is not massive enough to explode as a supernova (which is fortunate for us!) but it will eventually cast off its outer layers and produce a "planetary nebula"—a misleading name, because such an object has nothing to do with planets and is not truly a nebula. The best-known example is M.57, in Lyra, the Lyre, so-called because it was the 57th object in a catalog drawn up in 1781 by the French observer Charles Messier. Visually it looks like a tiny, shining cycle-tire, with a dim central star. Other planetary nebulae with picturesque nicknames such as the Cat's Eye, the Hourglass and the Helix are favorite Hubble objects; they are dying stars, and will eventually lose their outer shells altogether, turning into dim "white dwarfs." Our sun will become a white dwarf in several thousands of millions years' time, finally losing all its light and heat and becoming a cold, dead globe—a "black dwarf." However, this process takes an immensely long time, and our galaxy may not yet be old enough for any black dwarfs to have been formed.

At the other end of the time scale there are very young stars. The Orion Nebula, M.42, is easily visible with the naked eye, close to the three bright stars making up the Hunter's Belt; it is a patch of gas and dust, 1,500 light-years away, in which new stars are being born. Binoculars will show the multiple star Theta Orionis, known as the Trapezium, which shines upon the nebulosity and makes it glow. In Hubble images we see freshly-formed stars still with their original discs of dust, which are known as proplyds; it is within these proplyds that planets will gradually form. They are particularly well shown in the Orion Nebula because it is the closest of all the large nebula. Hubble has also given us an insight into brown dwarfs, which are stars that never began to radiate by nuclear fusion. I have likened them to stars which have failed their entrance examinations!

Brown dwarfs—a misleading name, because if we could see them from close range they would be red—have been under discussion for a long time, but are obviously difficult to find because of their low luminosity. Actually, the first to be found was originally detected with the 60 in (1.5 m) reflector at the Palomar Observatory in California, in October 1996, but was much more clearly seen by Hubble a year later, when its nature was confirmed. It is called Gliese 229B, because it is the

Opposite: The Orion Nebula (Hubble Space Telescope). This nebula, M.42, is easily seen with the naked eye; it is 1,500 light-years away, and is a stellar "nursery" in which new stars are being born from the nebular material. Note the stars of the Trapezium (Theta Orionis), on the Earth-turned side of the nebula, which excite it to luminosity.

faint companion of the red dwarf Gliese 229, 19 light-years away in the constellation of Lepus the Hare. It is thought to be around 30 times the mass of Jupiter, which is too massive for a planet and yet not massive enough to rank as a true star; it is a sort of missing link. It shines feebly because it is still contracting, and an astronaut unwise enough to land there would feel decidedly uncomfortable, but it has no source of energy other than that due to gravitational shrinking, and it will go on cooling down until the last of its light and heat leaves it. There must be vast numbers of them in the galaxy, both on their own or else, like Gliese 229B, members of binary systems.

It has been said that Jupiter is a kind of failed star, but this is not correct. For nuclear reactions to begin at its core, Jupiter would have to be at least eight times as massive as it actually is. A brown dwarf forms out of nebular material in the usual way, while a planet builds up from the material of a proplyd.

One of Hubble's most famous pictures shows what have been nicknamed the Pillars of Creation, in M.16, the Eagle Nebula in the Serpent, 7,000 light-years away. Here we have long columns rising from a huge cloud of cold hydrogen. On the ends of the columns there are EGGs—Evaporating Gaseous Globules. Both pillars and EGGs are being remorselessly eroded away by the flood of ultra-violet radiation sent out by energetic stars which already exist; this is termed photoevaporation. Forming inside the EGGs are a few stars which will eventually be uncovered.

Hubble has been a scientific triumph. It has allowed us to examine wonders which we could never have found without it. Inevitably it needs maintaining; quite apart from the first servicing mission, to correct for the original fault, there have been several others. The accountants are always prowling around, but at the time of writing, the telescope is still in full operation. What will finally happen remains unclear. To de-orbit it and allow it to disintegrate in the atmosphere would be heartbreaking, but at least there is no imminent crisis.

NEW SPACE TELESCOPES

For years after its launch the HST was in a class of its own so far as visual images were concerned—even though it has only a 94in (239cm) mirror. Now, in the twenty-first century, it no longer has the monopoly. For example, there is SIRTF, the Space Infrared Telescope Facility, sent aloft from Cape Canaveral on August 25, 2003; it was then renamed after Lyman Spitzer, an American astronomer who championed space research methods (perhaps because nobody knew quite how to pronounce "SIRTF"!). Its main work lies beyond the galaxy, but it may well have been able to reveal the specks of light from planets of other stars—which, as noted, are basically different from brown dwarfs. Another spectacular image shows the Double Helix Nebula, near the center of the galaxy; the shape may be due to ultra-strong magnetic fields generated by the gas disc orbiting the black hole, which lies in the exact center, 300 light-years from the nebula.

The Spitzer Space Telescope is scheduled to remain in orbit for about five years, but astronomers are hoping that it will last for longer.

Back with the solar system, we have TRACE, NASA's Transition Region and Coronal Explorer, one of a series of small space observatories. TRACE was launched into orbit to study the "transition region" of the sun's atmosphere, a dynamic region between the bright surface and the corona. The sun is never calm; there is ceaseless activity, and TRACE sends back images of violent storms and "loops" in the corona, most of which would contain many bodies the size of the earth.

All these investigations have been immensely successful. Yet our aim now is to look into the furthest depths of the universe, and this is where space telescopes are unrivaled.

Opposite: The "Pillars of Creation" in M.16, the Eagle Nebula in Serpens (the Serpent). This is one of the Hubble Space Telescope's most famous pictures, showing the columns and the EGGs (Evaporating Gaseous Globules). M.16, not far from the star Nu Ophiuchi (mag. 3.3) is of integrated magnitude 6.4; very small telescopes will show it as a misty patch.

Opposite: The Triangulum Spiral, M.33. It is the largest member of the Local Group after our galaxy and M.31, and the spiral form is well shown, because it is presented to us at a reasonably favourable angle. It contains objects of all kinds, including Cepheid variables.

11 THE DEPTHS OF THE UNIVERSE

Less than a century ago it was believed that the Milky Way galaxy made up the entire universe. It was known to be a flattened system, around 100,000 light-years across, and to contain at least a hundred thousand million stars, of which the sun was one. Around its edge, many thousands of light-years away, are the globular clusters—huge symmetrical systems, containing hundreds of thousands of stars each. Because they are so far away, not many of them are visible with the naked eye—the Hercules Cluster, M.13, in the northern hemisphere, and Omega Centauri and 47 Tucanae in the far south. Near their centers the stars are so close together that no Earth-based telescope can show them individually—though, of course, the stars are well separated and usually at least a light-year apart, and collisions are very rare indeed.

Gaseous nebulae, such as M.42 in Orion's sword, are common in the galaxy, but what about the "nebulae" that seemed to be starry? Opinions were divided. Some astronomers, notably Harlow Shapley (who had been the first to give an accurate value for the size of the Milky Way) believed them to be local and to be minor features. Others, such as Edwin Hubble, believed them to be independent galaxies, a very long way away.

Hubble won the argument when he discovered Cepheid variables in the Andromeda system. These convenient stars act as "standard candles" because their real luminosities depend upon their periods—that is to say, the interval between successive maxim; the longer the period, the more powerful the star. The prototype star, Delta Cephei (period 5.3 days) outshines Eta Aquilae in the Eagle (over 7 days). Once the luminosity is known, the distance can be worked out. Hubble measured the periods of the Andromeda Cepheids, and realized that they had to be much too remote to be members of the Milky Way. The Andromeda system, M.31, was after all a separate galaxy, and before the space telescopes came into operation the distance was given as 1,200,000 light-years.

(To recap, the M numbers for nebular objects were given by Charles Messier in his catalogue, published in 1781. NGC stands for New General Catalogue, although it is far from new; it was compiled in 1868 by J. L. Dreyer,

and later an "Index Catalogue" (IC) was added. In 1994, I listed over a hundred of the brightest objects not listed by Messier, and gave them "C" numbers. I never expected to be taken seriously, and was surprised when the C numbers became popular. By the way, C stands for Caldwell. Obviously I could not use M, and my surname is actually Caldwell-Moore. One sour amateur, annoyed because the idea had not been his, accused me of trying to spread my own name. I can only plead that this never occurred to me.)

It was expected that this value would be confirmed when, on August 8, 1989, the European Space Agency launched a new satellite, Hipparcos—High Precision Parallex Collector Satellite (the name recalls Hipparchus, the Greek astronomer who drew up the first good catalog around BC 140; Ptolemy's catalog was based on it). Hipparcos was an astrometric satellite, designed to measure the positions of 180,000 stars with unprecedented accuracy. It led to a new distance for M.31: 2.9 million light-years. However, more recent measurements reduce this to 1,500,000 light-years, so that we now see it as it used to be when the most intelligent creatures on Earth were jellyfish. (Sceptics viewing the news bulletins of the early twenty-first century have been known to suggest that this is still true.) Amazing pictures have been sent back from the Spitzer Space Telescope; the total number of stars in M.31 is around 10^{12} (that is to say, a million million), and the overall diameter is 220,000 light-years, so that M.31 is much larger and more populous than our Milky Way galaxy.

Galaxies occur in clusters or groups, and both our galaxy and M.31 belong to what is called the Local Group. Each group is receding from each other so that (as Hubble found out) the whole universe is expanding, but inside groups the motions are random, and at present M.31 and our galaxy are approaching each other at the rate of 300,000 mph (500,000 km/h). Eventually the two systems will collide, and although the stars are too widely spaced to hit each other (except very rarely) the interstellar dust and gas will be colliding all the time, triggering off intense star formation. Finally the two will merge, and the graceful spiral forms will be lost, to end up as a single giant elliptical system. The crisis is not imminent; the collision will not begin until about 5,000 million years in the future, by which time the earth—if it still exists—will be uninhabitable.

Another member of the Local Group is the Triangulum Spiral, M.33, rather smaller than Andromeda and rather further away (2,600,000 light-years). It is on the fringe of naked-eye visibility and binoculars show it easily. Recently it has been possible to measure its apparent proper motion; the rate at which it seems to shift against the background of more distant objects. A snail crawling on the surface of Mars would appear to move 100 times more quickly than the Triangulum galaxy.

We can see galaxy collisions going on. For example, Hubble has mapped the Antenae galaxies in Corvus, the Crow, NGC 4038, and NGC 4039 (0.60 and 61). They are 68,000,000 light-years away, and can be made out with a telescope of modest aperture. About a thousand million years ago they were separate systems, and both were spiral but from 600 million years ago they started to pass through each other, akin to orderly crowds walking in opposite directions. Stars were flung out, and by now the two streamers of these banished stars extend far beyond the original galaxies, looking a little like the antennae of an insect—hence the nickname. The MICE galaxies in Coma Berenices (NGC 4676, IC 825, and 820) are already colliding, preparatory to merging; their long tails are the result of tidy interaction. They are 300,000,000 light-years from us, and their spiral forms can still be made out. The Tadpole galaxy in Draco (the Dragon), no.188 in a catalog drawn up by the American astronomer Halton Arp, has a "tail" almost 300,000 light-years long, apparently produced by the pulp of a passing galaxy. The distance here is 420,000,000 light-years. Then we have the Cartwheel galaxy in Sculptor, with a ring-like structure 100,000 light-years across made up of very luminous young stars; the ring was due to a small intruder galaxy which passed through a larger system, compressing the interstellar material with the result that a "wave of star formation" rippled outwards from the impact point . The Cartwheel is a member of the Sculptor group of galaxies, about 500,00,000 light-years from us.

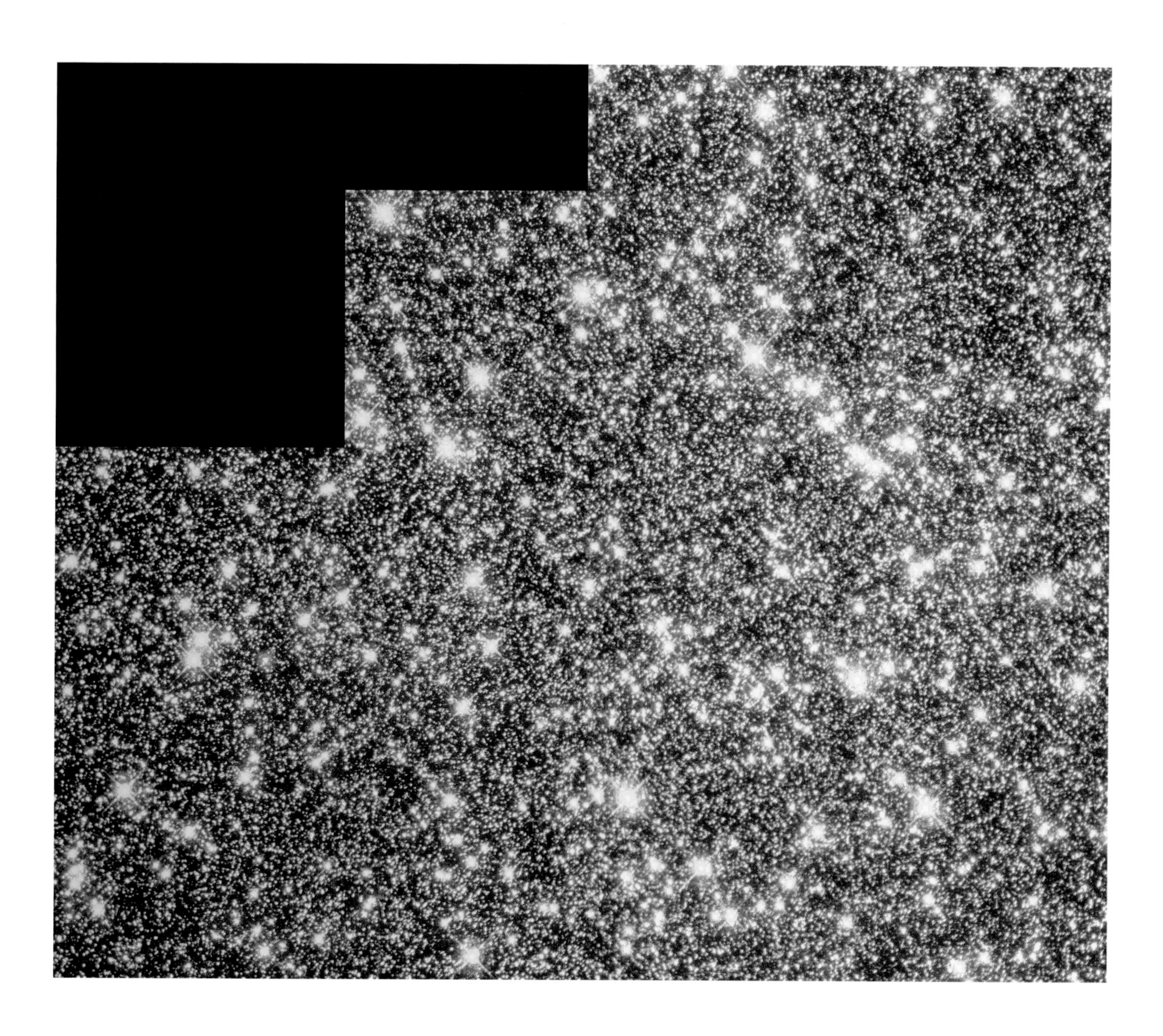

Opposite: Artist's impression of the *Hipparcos* spacecraft. *Hipparcos* (High Precision Parallax Collecting Satellite) was launched on August 8, 1989. It was to draw up a high-precision catalogue of 120,000 stars: the catalogue was published in 1997, but there have been suspicions that some of its data are not as reliable as had been hoped.

Above: Core of the globular cluster Omega Centauri, imaged by the Hubble Space Telescope. It is the largest of the globulars. It is 18,300 light-years away, and is thought to be 12 thousand million years old. This photograph includes around 50,000 stars. Omega Centauri may be the core of a small galaxy which was ripped apart and absorbed by our galaxy.

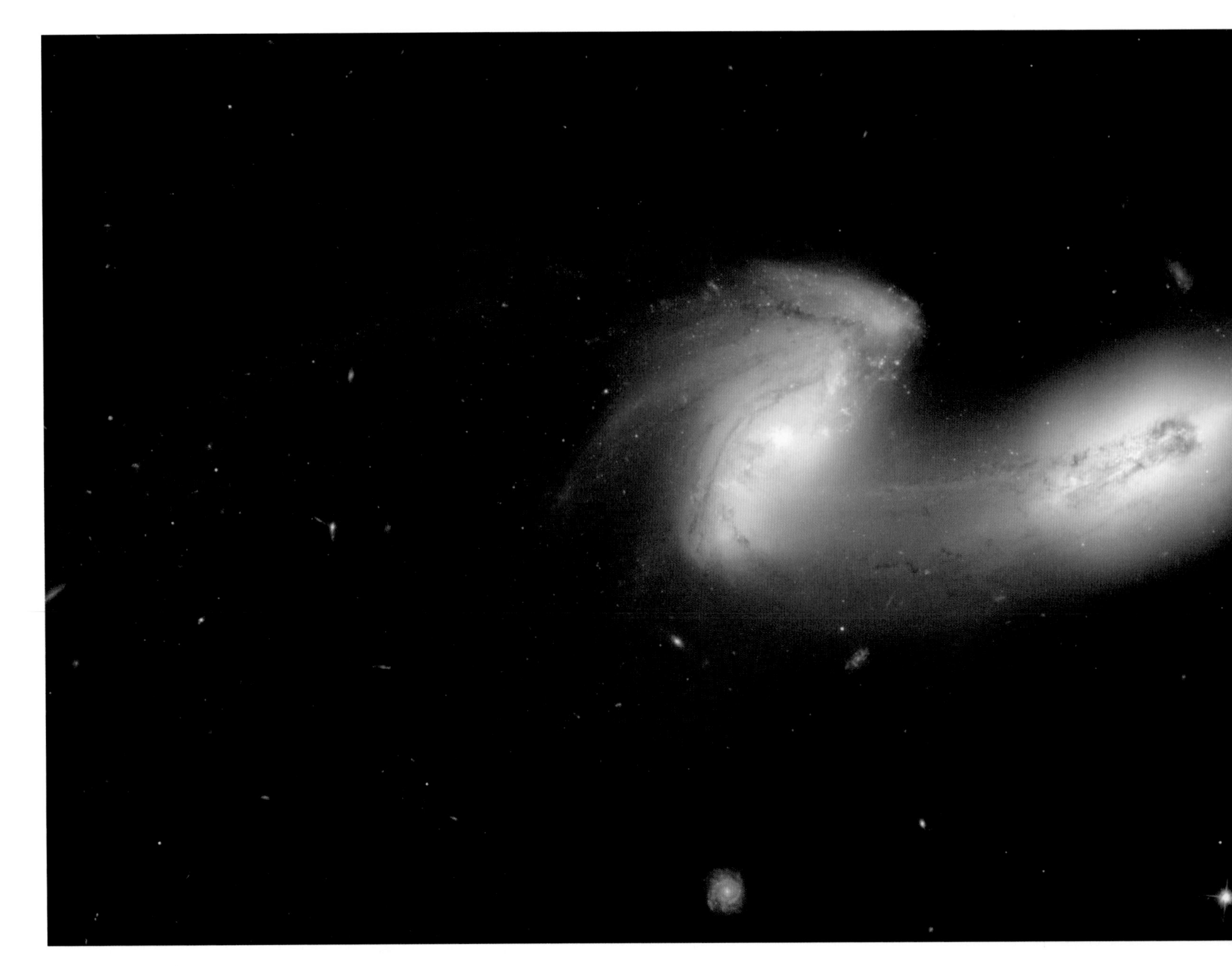

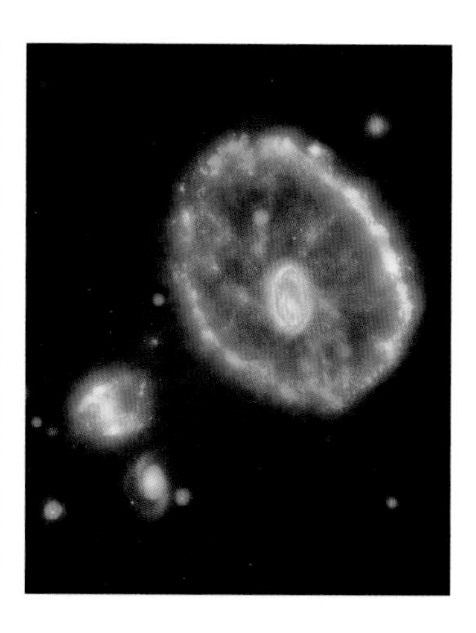

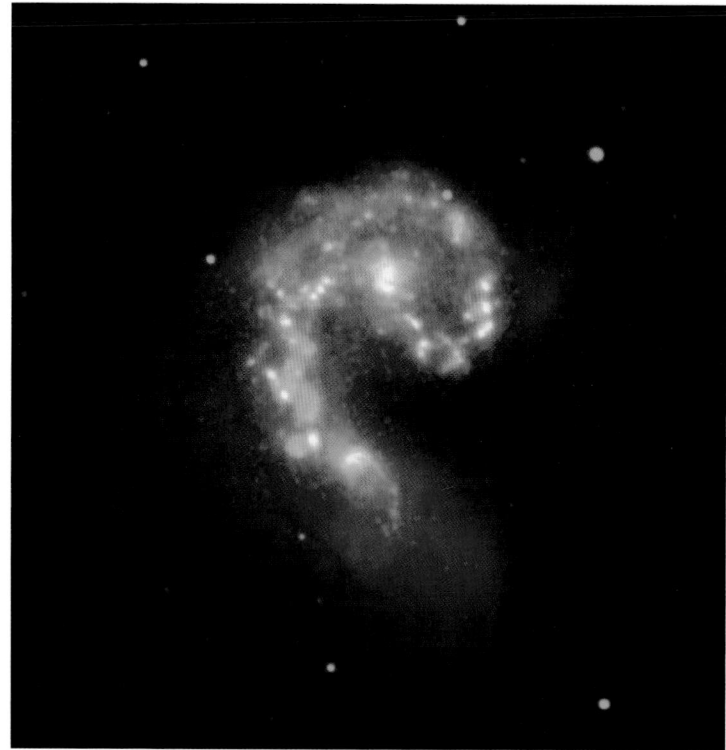

Collisions between galaxies are not brief affairs, and may take at least a thousand million years from first contact to final merger. Our Milky Way Galaxy has two satellite systems, the Large and Small Magellanic Clouds, at 169,000 and 210,000 light-years respectively. They are so-named because they were seen in 1519 by the great explorer Ferdinand Magellan, although they were known much earlier than that; the Arab Astronomer Al-Sufi recorded them as long ago as the year 964. (Northern observers always regret that they lie too far south in the sky to rise over any part of the Equator and cannot be seen in Europe.) They obtain stars of all kinds, together with open clusters, globular clusters and nebulae; in the Large Cloud we find the magnificent Tarantula Nebula, the most active star-forming region in the whole of the Local Group. In 1987 there was even one supernova, which became visible to the naked eye for some weeks, and temporarily altered the look of that part of the sky.

The clouds appear to be missing the Milky Way galaxy and are in no danger of disruption, at least at the present epoch, but this is not true of a small satellite system, the Sagittarius Dwarf Elliptica galaxy (SagDEG), which is now about 50,000 light-years from the galactic core,

Opposite: The Andromeda galaxy. Located 2.5 million light-years away, this is our largest nearby galactic neighbor. The galaxy's disc spans about 260,000 light-years, which means that a light beam would take 260,000 years to travel from one end of the galaxy to the other. The bright yellow spot at the centre is a particularly dense population of old stars.

Above: Colliding galaxies: NGC 4676A and B in Coma Berenices, nicknamed the Mice because of their long tails, produced by tidal action. They are 300 million light-years away, and belong to the Coma cluster. It is likely they will eventually merge.

Top left: NGC 4622, the "backwards" galaxy, seems to be rotating clockwise.

Top right: The Cartwheel galaxy, originally a normal galaxy, collided head-on with a smaller galaxy.

Above: The "Antennae" galaxies in Corvus.

and is on a direct collision course. Within the next hundred million years or so it will invade the inner part of our galaxy, and will be disrupted, so that it will cease to exist as a separate entity. It is only 80,000 light-years from the earth, but was not identified until 1994 because it lies on the far side of the centre of the Milky Way system.

Our galaxy is a spiral system, and like all spirals it is rotating; our sun takes 225 million years to complete one revolution around the galactic center. Predictably, the spiral arms are trailing. However, Hubble has imaged one galaxy, NGC 4622 in the Centaurus cluster (110,000,000 light-years away), which is behaving in a most extraordinary manner. It is rotating "backwards," so that the arms point forward—apart from the inner arms, which are conventional. NGC 4622 is a real enigma. It has been suggested that its unique rotation may be the result of a past mixed merger or a minor tidal encounter with a smaller companion galaxy, but no one really knows. The outer arms are characterized by bright bluish star clusters and dark dust lanes.

Above: The rapidly fading fireball from the most powerful cosmic explosion ever recorded—a gamma ray burst on January 23, 1999. For a brief instant the light from this blast was equal to the radiance of 100 million billion stars.

Opposite: Spitzer image of the Large Magellanic Cloud (LMC), 169,000 light-years away. It is a satellite of our galaxy, and contains objects of all kinds. It is easily visible to the naked eye. It was formerly believed to be an irregular system, but in fact seems to show indications of a barred spiral structure.

SPACE OBSERVATORIES

Space telescopes have proliferated in recent years. Many of the new "observatories" operate at definite parts of the electromagnetic spectrum. XMM-Newton (December 10, 1999) is the largest satellite ever launched by the European Space Agency; it is over 9 metres (30 feet) long and weighs 4 tons. It comprises two payload modules connected by a long carbon-fiber tube, which forms the optical telescope bench. X-ray telescopes, three of which are carried in XMM-Newton, are quite unlike optical telescopes, but are remarkably effective; for example Newton has discovered a vast ball of hot gas, 3,000,000 light-years across, streaking through a galaxy cluster known as Abell 3866, in the far southern constellations of Horologium (the Clock) and Reticulum (the Net). One member of the research team, Professor Mark Henriksen of Baltimore, said that "as the gas ball races through the cluster, individual galaxies tear portions of it away to stimulate future star growth," and he likened it to "a giant comet streaking through the heavens"; but a comet is icy, while the temperature of the gas ball is 100,000,000 °F (55,555,538 °C)—so hot that it cannot be seen in the optical range, but is striking afterglows at X-ray wavelength.

Then there is SWIFT, launched from Canaveral on November 20, 2004, and named after the world's fastest bird. Here the aim is to study gamma-ray bursts and their afterglows at the gamma ray, X-ray, ultraviolet and optical wavebands. Gamma rays are of very short wavelength—shorter even than X-rays—and GRBs, or gamma ray bursts, are the most powerful explosions known in the universe; in a few seconds, or at most a few minutes, a GRB emits a hundred thousand million times more energy than our sun does in a whole year. Studying them is difficult because they are so brief, and we never know when or where to expect them. SWIFT carries the Burst Alert Telescope (BAT) which covers a large fraction of the sky; when a burst occurs, BAT locates its position and relays it to ground stations, which in turn relay it to SWIFT. That done, SWIFT slows around to carry out detailed analyses. For a long time astronomers argued about the nature of GRBs, and it was even suggested that they might be fairly local, but we now know that a typical burst is caused by a very massive star collapsing to form a black hole. They are not rare, and within a few months of launch SWIFT

had recorded many of them. It is fortunate for us that they are so far away; if a GRB happened in our part of the Milky Way Galaxy, the results so far as we are concerned would be dire.

In 2005 SWIFT detected a GRB at a distance of 12.6 thousand million light-years. This means that the burst actually happened only a little more than a thousand million years after the birth of the universe in the original Big Bang, 13.7 thousand million years ago. This raises one fundamental question: how far can we see—and how far will we ever be able to see?

THE DEPTHS OF THE PAST

What is termed gravitational lensing helps. It was Einstein who realized that if the light from a very distant object passes by a massive intervening object, such as a galaxy, it will be bent so that its observed position may be different from its actual position; in fact, the intervening galaxy will act in the same manner as a lens. The story goes back to 1973, when astronomers at the Kitt Peak Observatory found what looked like a double quasar, cataloged now as 00957 561. Quasars are relatively small and intensely luminous; they are now known to be the cores of very active systems. But were the Kitt Peak astronomers seeing two separate objects? The answer was "no." The light from a single background quasar was being distorted by a galaxy between the quasar and ourselves.

If the source, the intervening object, and the earth lie in a straight line, the quasar will appear as a ring around the massive object. Hubble has obtained striking images of those "Einstein Rings." Gravitational lenses can also be used as what may be termed "gravitational telescopes," because they magnify objects lying behind them. In this way, Hubble, on February 15, 2004, detected the most remote galaxy so far known. We are looking back at the universe as it used to be at a very early stage in its existence.

We know that the groups of galaxies are racing away from us, and that the farther away they are the faster they are going. We must therefore come eventually to a distance where a

Left: Artists' impression of the XMM-Newton satellite, an orbiting X-ray observatory launched by the European Space Agency on December 10, 1999. It carries three X-ray telescopes plus various other instruments, and is very successful. Its distance from Earth ranges between 70,000 miles (112,654 km) at apogee and only 4,400 miles (7,081 km) at perigee.

Opposite: Astronomers were able to estimate from this Hubble Deep Field South image of invisible and infrared light that there are 125 billion galaxies above and beyond Earth's southern sky.

galaxy is receding at the full speed of light: 186,000 miles (299,338 kilometers) per second. We will then be unable to see it, and we will have reached the boundary of the observable universe, although not necessarily that of the universe itself. Working backwards, we find that the expansion started 13.7 thousand million years ago, which must therefore be the age of the universe in its present form. Just how the Big Bang happened is irrelevant in this context; what we want to know is—what were conditions immediately afterwards?

It is here that the Hubble Space Telescope is uniquely powerful, and in 1995 a new kind of experiment was carried out with it: the Hubble Deep Field (HDF). A small area of the sky was selected, measuring 144 arc seconds across, which is roughly the size of a tennis ball seen from a range of 300 ft. (100 m). The field was in the constellation of Ursa Major (the Great Bear), well away from the Milky Way and its obscuring clouds of gas and dust; only a few foreground stars were present, and nothing much else—in fact the field was chosen because it seemed to contain absolutely no objects of note. Between December 18 and 28, the Wide Field and Planetary Camera took 342 separate exposures of it, and combined them. The results were fascinating. About 3,000 distinct galaxies were shown, some spiral and some irregular, together with 50 blue point-like objects, some of which will probably be quasars while others are thought to be regions of intense star formation.

It was found that the HDF galaxies contained a higher proportion of irregular and disturbed galaxies than we find elsewhere; collisions and mergers were commoner when the universe was young. Studies of galaxies at different stages of their evolution have led to estimates over the rate of star formation over the ages. It may possibly have peaked from 8 to 10 thousand million years ago, and has since declined. Some astronomers have even suggested that the universe has begun its "long twilight" before death…

Three years after the HDF experiment a second Deep Field was selected, this time in the southern hemisphere of the sky. The results were much the same.

Next came the even more ambitious Hubble Ultra Deep Field (HUDF); the observations began on September 24, 2003, and went on until January 16, 2004, using the very latest equipment including the "grism" spectograph, a hybrid prism, and diffraction grating. Galaxies could be seen that were three or four times fainter than any Hubble could show before. The target area was in the southern constellation of Fornax, the Furnace, southwest of Orion. Altogether 800 exposures were taken, during which Hubble went round the earth 400 times.

HUDF takes us back over 13,000 million years, to a time when the first galaxies were starting to appear after the end of the long "Dark Ages" when the universe was opaque. One astronomer—Massini Stiaveklin of the Space Telescope Science Institute at Baltimore—commented that, "whereas the HDFs showed galaxies when they were youngsters, HUDF reveals them as toddlers, enmeshed in a period of rapid developmental changes."

Even so, we are not yet back to the time of the Big Bang, and we have no choice but to speculate. We cannot say "where" the Big Bang happened, because if space, time, and matter came into existence at the same moment as modern theory dictates, the Big Bang happened "everywhere." It is clear, then, that the universe has no "center." Similarly, if time began with the Big Bang, there was no "before." Concepts of this kind are really impossible to imagine, but the alternatives leave us no better off.

In 1948 Hermann Pondi and Thomas Gold, soon supported by Fred Hoyle, proposed the Steady-State theory of the universe. There was no Big Bang; the universe has always existed, and will exist forever. As old galaxies die, they are replaced by new ones, created spontaneously out of nothingness. The rate of creation would be too slow to be measured, but the theory laid down that the universe has always looked much the same as it does now. Results from the space telescopes, in particular Hubble's HDF and HUDF, prove that this is not so; the young universe looked quite different from the regions we know. In any case, we are replacing one impossible concept with another, because we just cannot imagine a period of time that has no beginning.

The one inescapable fact is that you, me, this book and everything else actually exist. The material making it up must have been created somehow or other, but to solve these fundamental problems we must agree that our brains are several dimensions short. Twenty-first century science, even space science, may have taken us as far as it can.

Top left: **2001**
On September 11, a series of coordinated terrorist attacks are launched on the U.S. by Islamist extremists affiliated with Al Qaeda. New York's World Trade Center is hit by two hijacked commercial passenger jetliners. A third airliner is crashed into the Pentagon near Washington DC. A fourth hijacked flight crashes into a field in Pennsylvania. In addition to the 19 hijackers, 2,973 people die.

Bottom left: **2002**
January 1 – In the biggest monetary changeover in history, Euro notes and coins enter circulation in 12 European Union countries.

Top right: **2001**
America's seventh largest company, the energy firm Enron, collapses in one of the country's largest ever corporate scandals. The company achieves infamy when it is revealed that its success involved a series of elaborate and fraudulant financial scams.

Bottom right: **2003**
The 2003 Iraq War begins with the invasion and occupation of Baghdad by a U.S.-led coalition force without UN support. The objectives of the occupation are to disarm the country of "weapons of mass destruction," end Saddam Hussein's support for terrorism, and free the Iraqi people. Protesters and anti-war activists stage demonstrations around the world against what is described as an illegal war.

2000–2007

Top left: **2004**
December 26 – Massive sea surges triggered by an earthquake under the Indian Ocean kill approximately 300,000 people in southern Asia. An 8.9 magnitude earthquake under the sea near Aceh, north Indonesia, generated the biggest tsunami the world had seen for at least 40 years.

Top right: **2005**
February 16 – After years of delays, the 141-nation Kyoto Protocol formally takes effect. The world plan to fight global warming by reducing greenhouse gas emissions is celebrated by its supporters as a lifeline for the planet but rejected as an economic straitjacket by the United States and Australia, who refuse to sign up to the treaty.

Bottom left: **2005**
August 29 – Devastation occurs when Hurricane Katrina makes landfall on America's Gulf Coast. Katrina was the costliest and one of the deadliest hurricanes in the history of the United States. The catastrophic effects of destruction and flooding were most notable in the city of New Orleans, Louisiana, and in coastal Mississippi.

Opposite: NASA's *New Horizons* probe launching in 2006. Its mission is to visit Pluto and its moons in 2015, shedding light on their surface characteristics, geology, interiors, and atmospheres. To save three years' flight time on its way to the edge of the solar system, *New Horizons* used Jupiter's gravity like a slingshot, increasing its speed by 2.5 miles per second (4 kilometers per second).

12
THE NEXT 50 YEARS

The last half-century has been a period of pioneering. The space age has started, and is well under way; artificial satellites have been succeeded by space stations; unmanned probes have been sent past all the planets; telescopes and other instruments of unrivaled power have been deployed, and, of course, men have landed on the moon. So what do the next 50 years have in store for us?

We have to admit that everything depends upon the world situation. As I write these words in 2007 one major war is in full swing, in the Middle East, and there are any number of less-publicized skirmishes. One more global conflict will put paid to not only space exploration, but, quite possibly, the whole of civilization. We may well look towards other stars that seem to be suitable candidates as planetary centers—for example Epsilon Eridani, a mere 11 light-years away. There are a hundred thousand million stars in our galaxy, and many of them have planets moving around them; on an Earth-like planet orbiting a sun-like star, it is not unreasonable to expect Earth-like life. If other civilizations have developed in the same way as ours, they may have far outstripped us in knowledge and culture, but on the other hand may have built "weapons of mass destruction" and used them to wipe each other out in the manner of Kilkenny cats, leaving a ruined and uninhabitable planet. We can only hope that Homo sapiens do not act in such a way.

Any other civilization must be light-years away, and indeed Epsilon Eridani, an undistinguished-looking star considerably less luminous than our sun, may be the most promising of our stellar neighbors. Sending a spacecraft there is hopelessly beyond us, but there is always the chance of establishing contact by radio, and serious attempts have been going on to pick up non-natural signals from afar. So far the results have been negative, but there have been several false alarms (one cosmic signal turned out to be due to the electric oven in the observatory kitchen!) but the quest is not pointless. There is a miniscule but definite chance of success.

NEW HORIZONS
NASA
LOCKHEED MARTIN
ILS

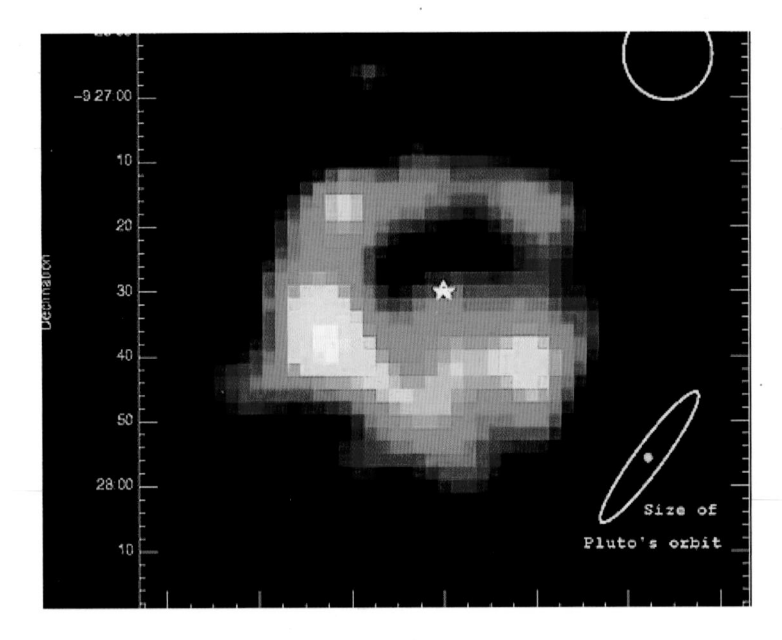

The chances that we will be visited by any aliens are even more miniscule. No doubt there will be many more UFO sightings, more Roswells, and more crop circles, but this is only to be expected, and we would all miss the UFOlogists, the Flat Earth supporters and their kind, to say nothing of the super-eccentrics who solemnly believe that the Apollo landings were faked in a NASA studio.

Unquestionably there will be many more unmanned space observatories. The Hubble Space Telescope has already lasted for longer than its originally planned lifetime, and is still working almost as well as ever, but it cannot operate indefinitely, and it will have to be replaced. Bringing it back to Earth, giving it a thorough update and overhaul is—alas—not practical, so what will eventually happen to it? To de-orbit it and leave it to burn away during descent would be heartbreaking; the obvious course is to "boost" it into a higher orbit, well above the last traces of the resisting atmosphere, and leave it there until our technology has developed sufficiently to cope with it. Its planned successor, the James Webb Space Telescope, will be much larger, but will operate mainly in the infrared. Also, it will be stationed at one of the Lagrangian points, over a million kilometres away from Earth. This will be an ideal place observationally, well away from terrestrial pollution, but no Hubble-type servicing missions will be possible. Hubble had to be repaired immediately after launch; with the JWST, one can only hope that the constructors will get everything right first time.

Launch of the James Webb Space Telescope was originally scheduled for around 2009. Then the accountants stepped in, and put the date back to 2012. They have since forced a further postponement, to 2013, and even this is under threat. Presumably we must wait to see how many extra dollars are required for wars in Iraq, Afghanistan, and elsewhere. (The total cost of these adventures would have financed space research well into the twenty-second century—perhaps even into the twenty-third century.) But we must hope for the best.

The space station itself has widely been criticized as being the cosmic equivalent of a white elephant, but scientifically it has already proved its worth, and has the added advantage of being genuinely international. It is not yet complete, but surely it cannot now be abandoned. But let us now look further ahead, to AD 2057, and see what we may have achieved.

What about space tourism? It has already started, and several multi-millionaires have been taken into orbit. In the foreseeable future other tourists will follow, not all of them equally wealthy, and it may well be that by 2057 a trip to the space station will be no more difficult than a journey to the South Pole used to be in 1957. Yet we must always remember that space remains a hostile environment, and occasional tragedies are only to be expected.

A fully-fledged lunar base is a real possibility. If financed, it could probably be in place by 2030 or earlier. Initially it would be manned exclusively by scientists, and there would be some aspects of special importance. Medical research will take priority, because the moon, with its weak gravity and lack of atmosphere, provides an environment which cannot be simulated on Earth. A "lunar hospital" is likely to prove of immense value to all humanity. Patients will be treated there—always provided that they can tolerate the journey from Earth.

MISSIONS TO MARS

Next on the target list comes Mars, the only world within reasonable range that we can possibly hope to reach before 2050 (we can forget

Above: The star Epsilon Eridani is surrounded by a ring of dust, amongst which is a Jupiter-like planet. As our nearest extra-solar planet, it is an inevitable focus for speculation about other life in the universe.

Opposite: An artist's impression of NASA's Crew Exploration and Lunar Lander Vehicles, being developed to resume manned missions to the moon. Capable of carrying six people instead of *Apollo*'s three, it has been described as "*Apollo* on steroids."

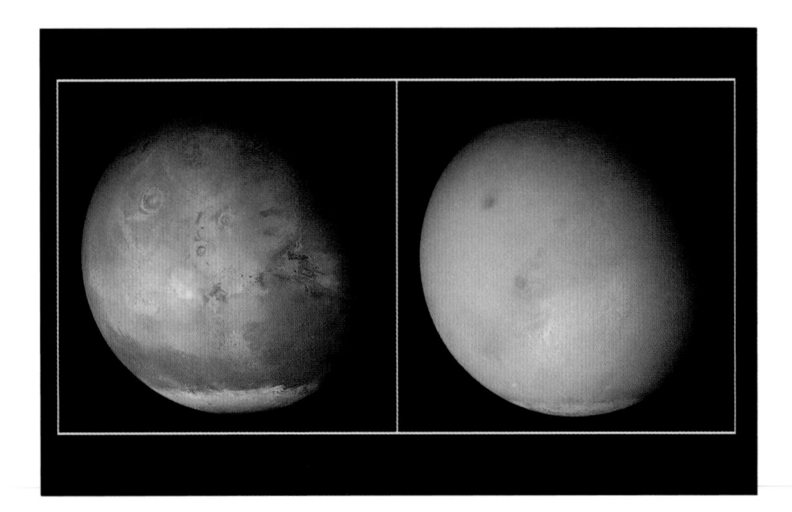

about Venus and Mercury so far as manned expeditions are concerned, at least for the moment). We have to accept that the problems are many, many times more pressing than those of a flight to the moon. First, there is the journey itself, which will last for many weeks. Radiation is potentially deadly, and not easy to block. If a violent solar storm occurs, the astronauts will be very much at risk, and in my view this is the one thing that may cause indefinite delay. Quite apart from this, we are by no means sure how the human body will cope. True, Russian cosmonauts have remained in orbit, under conditions of weightlessness, for over a year at a time, but circling the earth in the outer atmosphere is very different from undertaking a journey millions of kilometres away, and on reaching Mars there will be no hospitals ready to deal with any casualties.

Before planning an expedition, we must know a great deal more about Mars than we do at present. The landers, particularly the rovers, have assured us that the surface is pleasingly solid, and though dust storms are frequent they will not pose serious threats; the winds are fast, but in that tenuous atmosphere they will have relatively little force. The most they can really do is to reduce visibility. Neither will "marsquakes" be strong enough to be worrying. In all these respects Mars will be safe—but what about the levels of radiation on the surface?

The atmospheric pressure is below 10 millibars everywhere, so from our point of view, the "air" would be useless even if it were made of pure oxygen; but it is not—it is made up almost entirely of carbon dioxide. It is extensive enough, but whether it will provide an adequate radiation screen is another matter, and at present we simply do not know. There is much to be done before giving any date for a Martian adventure. Over the next decade a large number of robot probes will certainly be dispatched, and with luck will tell us how serious the problems really are.

There is one extra question: Can we expect to find anything dangerous on Mars? Here the answer is almost certainly "no," but I stress the "almost." Mars, unlike the moon, is not inert. There is an atmosphere; there may still be a certain amount of volcanic activity, because the great volcanoes may be dormant rather than extinct; there was probably life in the remote past, and lowly organisms may survive today. When the *Apollo* astronauts came back from the moon, they were kept in strict isolation until it was definite that they had brought back nothing malignant. Quarantining was abandoned after *Apollo 12*, because it was clear that the danger was nil, but with Mars we are faced with an entirely new situation, and we cannot afford to take any risks. In all probability the returning astronauts will remain in orbit until they have been pronounced "safe."

Before a manned expedition sets off, there will have been many robot spacecraft. Samples of Mars will be brought back to Earth for analysis, and this is likely to happen within the next few years. Materials and supplies will be deposited on the surface to await the arrival of astronauts; remember, the pioneers will have to stay for months, so there is no comparison with a "there-and-back" flip to the moon. At least there will be no shortage of water, or rather H_20 in the form of ice.

The pioneer expedition will probably include several spacecraft. There is a strong chance that the initial blastoff will be from the lunar base, and personally I believe that the first stop may not be Mars, but one of the two satellites. These two midgets, Phobos and Deimos, were discovered in 1877, and are quite unlike our massive moon. They are irregular in shape; Phobos has a longest diameter of less than 20 miles (32 km), Deimos less than 10 miles (16 km); without doubt they are ex-asteroids, captured by Mars long ago. Neither would be of any real use as sources of illumination at night, and indeed Deimos would appear as little more than a tiny disc. Phobos is so close in that it completes one orbit in only seven and a half hours—less than a Martian day, or "sol," which is just over half an hour longer than ours. An observer on Mars would see Phobos rise in the west, gallop across the sky, and set in the east four and a half hours later; this would happen three times in every sol. Deimos is much further out from Mars (12,000 miles/19,312 km) and would remain above the horizon for two and a half sols consecutively. It may be best to touch down first on one of these two, preferably Deimos, and use it as a space station, as though it had been put there specially for our benefit! Of course, the gravitational pull of Deimos is so weak that rendezvous will be in the nature of a docking operation rather than a true landing.

It is too early to say whether a permanent Martian base will be established; there are so many factors to be taken into account, notably the extent of the radiation hazard. The idea of "terraforming" Mars, and

Opposite: The image on the right shows a dust storm brewing on Mars in June 2001; the image on the left shows the planet almost entirely engulfed three months later. Although dust storms are frequent and dramatic on Mars, they are unlikely to hinder our exploration of the planet.

Below: The first-ever space tourist, multimillionaire Dennis Tito (center) gives a press conference on board the *International Space Station* in 2001. Several others have followed in his footsteps, and within the next 50 years, space tourism could become routine.

making it into a sort of second earth, appeals to visionaries such as Arthur C. Clarke, but if this is ever attempted we are looking at centuries ahead.

FUTURE HORIZONS

So far as named landings are concerned over the next 50 years, we are limited to the moon and perhaps Mars, but there will certainly be many unpiloted probes to all the planets. Controlled landings on Mercury are feasible, and instruments there will be ideally placed for studying the sun as well as comets that venture into these torrid regions—and some of which do not survive. Venus is less promising, and even in 2057 the best means of improving our knowledge may still be by radar equipment carried in orbiters. There will no doubt be encounters with asteroids, and if one of these cosmic mavericks is found to be on a collision course efforts may be made to divert it—not by shattering it, but by exploding a nuclear weapon close beside it and "nudging" it clear of Earth. This depends upon whether we've had enough warning, and if the asteroid happened to be several kilometres in diameter there would not be a great deal that we could do. Fortunately, the chances of a major collision before 2057 are very slight indeed.

The outer planets will not be neglected. One probe, New Horizons, is already on its way to Pluto, and will arrive there in 2015; we now know that Pluto is merely the brightest (not the largest) of the Kuiper Belt objects, but it is still of great interest. So of course are the giant planets—and their satellites. In Jupiter's family, Io has wildly active volcanoes, erupting all the time. Europa has an icy surface, below which there may well be an extensive ocean of ordinary water. It is hard to believe that any living thing could survive in a pitch-dark, sunless sea—but one never knows, and life does exist in the most unexpected places, such as undersea hydrothermal vents. Saturn's largest attendant, Titan, has methane rivers; the much smaller Enceladus spouts ice crystals from below its surface; Triton, the only large satellite of Neptune, presents nitrogen geysers—all these worlds are different, and all have much to tell us. Now, for the first time in the history of mankind, we have the ability to answer many of the questions they ask us.

But we must always remember that in paying so much attention to the solar system, we are being parochial. Our earth is unimportant in the grand scheme of things; so is the sun; so is the Milky Way Galaxy; so is the local group of galaxies. What we aim to do, eventually, is to gain a real understanding of the universe. It will take us a long time, if we ever succeed, but when we look back at what we have achieved in relatively recent times we are entitled to be cautiously optimistic. In 1807 we knew that the stars were suns many light-years away, but we had little idea of the extent of the universe itself; we did not know how the sun shines, and we did not know the age of the earth. By 1907 we had a better idea of the scale of the universe, but we still believed our galaxy to be the only one. We had—just—learned to fly in heavier-than-air machines, but any suggestion of travelling to another world, even the nearby Moon, was dismissed as unworthy of serious discussion. Astronomy was limited to the wavelengths of visible light; even radio astronomy lay in the future. The largest telescope in the world was still Lord Rosse's homemade 72 in. (183 cm) reflector, with its metal mirror.

Fifty years later, in 1957, the situation had changed. Edwin Hubble had proved that our galaxy is one of myriads, and that the entire universe is expanding; we can peer millions of light-years into space. Powerful telescopes had come upon the scene, notably the 200 in. (5.1 m) reflector at Palomar Observatory in California. Invisible astronomy was under way, with the 250 ft. (76 m) radio telescope at Jodrell Bank. Rockets had been developed by pioneers such as Wernher von Braun, and the moon at least was accepted as being within range. Members of Interplanetary Societies (such as myself) were no longer regarded as amiable lunatics. Yet there were still people who remained convinced that practical space research would never be possible.

This was a mere half-century ago. With Sputnik 1, in October 1957, all doubts were swept aside. Our task is now to build upon what we have achieved, and to make sure that we work for the benefit of mankind rather than using our expertise to build real "weapons of mass destruction." We are well and truly into the space age, and we live in the most exciting period of human history.

REFER

RENCE

ALTAZIMUTH MOUNT
A type of telescope mounting which allows the instrument to move freely in any direction.

ALTITUDE
The angular distance of a celestial body above the horizon.

ANGSTROM UNIT
A hundred-millionth of a centimeter.

APHELION
The orbital position of a planet or other body when at its greatest distance from the sun.

CASSINI DIVISION
The main division in Saturn's ring system (between Rings A and B).

CELESTIAL SPHERE
An imaginary sphere surrounding the earth, whose center coincides with that of the earth.

CEPHEID
A short-period variable star, whose period is linked with its real luminosity. The name comes from the prototype star, Delta Cephei.

CHROMOSPHERE
The layer of the sun's atmosphere lying above the bright surface

CONJUNCTION
There are several meanings. A planet is said to be in conjunction with a star when it passes close by in they sky. Mercury and Venus are said to be at inferior conjunction when exactly between the earth and the sun, and at superior conjunction when lined up with earth on the far side of the sun. The remaining planets pass through superior conjunction only.

CONSTELLATION
A group of unassociated stars, arbitrarily named. The International Astronomical Union officially recognizes 88 constellations.

CORONA
The outermost part of the sun's atmosphere, made up of very tenuous gas at a very high temperature (though because it is so rarefied, there is very little "heat").

COSMIC RAYS
High-velocity atomic particles reaching the earth from outer space. The heaviest (primary) particles are broken up on entering the atmosphere—which is fortunate.

DECLINATION
The angular distance of a celestial body north or south of the celestial equator. Declination in the sky corresponds to latitude on Earth.

DEGREE OF ARC
A unit for measuring angles. A full circle contains 360 degrees; each degree is divided into 60 minutes, and each minute into 60 seconds.

DICHOTOMY
The exact half-phase of Mercury, Venus, or the moon.

DIRECT MOTION
Bodies which move around the sun in the same sense as the earth. The term also applies to planetary satellites; a satellite has direct motion when it orbits in the same sense as the planet's rotation. Movement in the opposite sense is termed retrograde.

DIURNAL MOTION
The apparent rotation of the sky from east to west due to the earth's real rotation from west to east.

DOUBLE STAR
A star made up of two components. Optical doubles are due to chance alignment as seen from Earth; the components of a binary are genuinely associated.

DWARF NOVAE
A term sometimes applied to U Geminorum variable stars.

EARTHSHINE
The dim visibility of the non-sunlit side of the crescent moon, due to light reflected on to the moon from the earth.

ECLIPSES, LUNAR
These are produced when the moon enters the cone of shadow cast by the earth. A lunar eclipse may be either total or partial.

ECLIPSES, SOLAR
Produced when the new moon passes in front of the sun. A solar eclipse may be total, partial or, annular; at an annular eclipse the moon is in the far part of its orbit, and does not look large enough to cover the sun completely—so that a ring of the solar photosphere is left showing around the dark disc of the moon.

ECLIPSING VARIABLE (ECLIPSING BINARY)
A binary star in which the components periodically pass in front of each other and cause apparent variations in light. The most famous eclipsing variable is Algol in Perseus.

ECLIPTIC
The projection of the earth's orbit on to the celestial sphere. It may also be defined as the apparent yearly path of the sun against the stars. It is inclined to the celestial equator by an angle of about 23½ degrees.

ELECTROMAGNETIC SPECTRUM
The full range of wavelengths—from the very long radio waves through to the ultra-short gamma-rays.

ELEMENT
A substance which cannot be chemically split up into simpler substances.

ELONGATION
The apparent angular distance of a celestial body from the sun, or of a satellite from its parent planet.

EPOCH
A date chosen for reference purposes in quoting astronomical data.

EQUATOR, CELESTIAL
The projection of the earth's equator on to the celestial sphere.

EQUATORIAL MOUNT
A telescope mounted upon an axis which is parallel to the axis of the earth so that only east-to-west movement is needed to keep the target object in the field of view.

EQUINOXES
The two points at which the ecliptic cuts the celestial equator.

ESCAPE VELOCITY
The minimum velocity at which an object must move in order to escape from the surface of a planet, or other celestial body, without being given extra impetus.

EVENT HORIZON
The boundary of a black hole. No light can escape from inside.

EXOSPHERE
The outermost part of the earth's atmosphere.

EXTINCTION
The apparent dimming of a star or planet when low in the sky. It is due to absorption of light in the earth's atmosphere.

EYEPIECE (OR OCULAR)
The lens, or combination of lenses, at the eye-end of a telescope responsible for magnifying the image of the object under study.

FACULAE
Bright temporary patches on the surface of the sun.

FIREBALL
A very brilliant meteor (conventionally, above around magnitude-5).

FLARES, SOLAR
Brilliant eruptions in the outer part of the sun's atmosphere. They emit charged particles which, on reaching the earth, cause magnetic storms and displays of aurorae.

FRAUNHOFER LINES
Dark absorption lines in the spectrum of the sun.

GALAXIES
Systems made up of stars, nebulae, and interstellar matter—plus a great quantity of dark matter, which we cannot see.

GALAXY, THE
Our own galaxy; it is a rather loose spiral, containing 100 thousand million stars.

GAMMA-RAYS
Radiation of extremely short wavelength.

GIBBOUS PHASE
The phase of the moon or a planet between half and full.

GLOBULES
Small dark patches inside gaseous nebulae.

HELIOSPHERE
The region round the sun inside which the sun's influence is dominant.

HERTZSPRUNG–RUSSELL DIAGRAM (H–R DIAGRAM)
A diagram in which stars are plotted to their spectral types and their luminosities.

HUBBLE CONSTANT
A measure of the rate of recession of a galaxy. It is given as 70 kilometers per second per megaparsec.

KUIPER BELT
A belt of asteroid-sized objects beyond the orbit of Neptune.

LIBRATION
The apparent "swaying" of the moon as seen from the earth. There are several librations, and all in all we can see 59 percent of the total surface, though of course no more than 50 percent at any one time.

LIGHT-YEAR
The distance traveled by light in one year: 5,879,000,000,000 miles (9.46 million million kilometers).

LOCAL GROUP
The group of galaxies of which the Milky Way Galaxy is a member. The other major systems are the Andromeda and Triangulum Spirals.

LOCAL SUPERCLUSTER
A supercluster of galaxies, centered on the Virgo cluster and including our Local Group.

LUNATION
The interval between successive new moons: 29 days, 12 hours, 44 minutes.

MAIN SEQUENCE
A band across the H–R Diagram, including the vast majority of stars.

MEGAPARSEC
One million parsecs.

MERIDIAN, CELESTIAL
The great circle on the celestial sphere, which passes through the zenith and both celestial poles.

MESSIER CATALOGUE
A catalogue of 109 prominent star clusters and nebulae.

METEOR
A small piece of cometary debris, which dashes into the earth's upper atmosphere and burns away by friction.

METEORITE
A larger body than a meteor, which lands upon earth without being destroyed. Meteorites come from the asteroid belt, and are not associated with meteors.

MICRON
A thousandth of a millimeter.

MOLECULAR CLOUD
A cloud of interstellar matter in which the gas is mainly in molecular form.

NADIR
The point on the celestial sphere directly below the observer.

NEBULA
A cloud of gas and dust in space.

NEUTRINO
A fundamental particle with no electric charge and negligible mass.

NEUTRON STAR
The remnant of a massive star that has exploded as a supernova.

NOVA
The white dwarf component of a binary system which suffers a violent but temporary outburst. It then reverses to its former brightness.

OBLIQUITY OF THE ECLIPTIC
The angle between the ecliptic and the plane of the earth's orbit; 23° 26′ 45″.

PARALLAX, TRIGONOMETRICAL
The apparent shift of an object when observed from two different directions.

PARSEC
The distance at which a star would have an annual parallax of 1 second of arc: it is equal to 3.26 light-years.

PENUMBRA
(1) The area of partial shadow to either side of the main cone of shadow cast by the earth. (2) The outer, lighter part of a sunspot.

PERIGEE
The position of the moon in its orbit when closest to the earth.

PERIHELION
The position of a planet or other body in its orbit when closest to the sun.

PHASES
The apparent changes in the shape of the moon from new to full. Mercury and Venus show lunar-type phases, while Mars can be appreciably gibbous.

PHOTOSPHERE
The bright surface of the sun.

POLES, CELESTIAL
The north and south poles of the celestial sphere.

PRECESSION
The apparent slow movement of the celestial poles. It is due to the pulls of the sun and the moon on the earth's equatorial bulge.

PRIME MERIDIAN
The meridian on the earth's surface which passes through the Airy transit circle at Greenwich Observatory. It marks longitude 0°.

PROMINENCES
Masses of glowing gas rising from the sun's surface; they are composed mainly of hydrogen.

PROPER MOTIONS
The individual movement of a star on the celestial sphere.

PULSAR
A rotating neutron star sending out beams of radio radiation.

QUADRATURE
The position of the moon or a planet when at right angles to the sun as seen from the earth.

RADIAL VELOCITY
The movement of a celestial body towards or away from the earth: positive if receding, negative if approaching.

RADIANT
The point in the sky from which the meteors of any particular shower appear to radiate.

RETROGRADE MOTION
Orbital or rotational motion in the sense opposite to that of the earth, or, in the case of a satellite, the sense of the primary planet's rotation.

RIGHT ASCENSION
The angular distance of a celestial body from the vernal equinox, measured eastward. It is usually given in units of time, so that the RA of the body is the difference between the culmination of the vernal equinox of the culmination of the body.

SOLSTICES
The times when the sun is at its maximum declination (23½ degrees) from the celestial equator.

SPECIFIC GRAVITY
The density of a substance, taking that of water as 1.

STARBURST GALAXY
A galaxy showing an exceptionally high rate of star formation.

SUPERIOR PLANETS
All planets whose orbits lie outside that of the earth (i.e., all planets except Mercury and Venus).

SUPERNOVA
A colossal stellar outburst, involving either (1) the collapse of a very massive star, or (2) the total destruction of the white dwarf component of a binary pair.

SYNODIC PERIOD
The interval between the successive oppositions of a superior planet.

TERMINATION
The boundary between the day and night hemispheres of the moon or other planetary body.

TRANSIT
(1) The passage of a celestial body across the observer's meridian. (2) The projection of Mercury or Venus against the sun, or of a satellite against its primary planet.

UMBRA
(1) The main cone of shadow cast by the earth. (2) The dark inner part of a sunspot.

WHITE DWARF
A very small, very dense star which has used up all its nuclear "fuel."

ZENITH
The observer's overhead point (altitude 90°).

ZODIAC
The belt around the sky 8° on either side of the ecliptic, in which the sun, moon, and planets are always to be found.

ZODIACAL LIGHT
A cone of light rising from the horizon and extending along the ecliptic. It is due to sunlight striking thinly-spread material near the main plane of the solar system.

INDEX

Figures in italics indicate captions; those in bold indicate glossary terms.

ACKNOWLEDGMENTS

The publisher would like to thank Andrew Ellard and Alec Edgington for their help with this project.

Picture Credits

Key: T= Top; B= Bottom; M = Middle; TR = Top right; TL = Top left, etc.

2: NASA/JPL ; 3: NASA/JPL; 7: NASA/JPL; 9: NASA/JPL; 10 L: Ria Novosti/Science Photo Library; 10R: Stephano Bianchetti/Corbis; 11 T: Gustavo Tomsich/Corbis; 11 B: Yerkes Observatory; 12: Roger Ressmeyer/Corbis; 13: David Parker/Science Photo Library; 14: The International Astronomical Union/ Martin Kornmesser; 15: The Institute for Solar Physics, Sweden; 16 TL: Bert Hardy/Picture Post/Getty Images; 16 TR: Keystone/Getty Images; 16 BL: Al Fenn/Time Life Pictures/Getty Images; 16 BR: CBS Photo Achive/Getty Images; 17T: OFF/AFP/Getty Images; 17 BL: Keystone/Getty Images; 17BR: Hulton Archive/Getty Images; 19: Sol Invictus/NASA; 21: Corbis; 22: Sol Invictus; 23: Sol Invictus; 24: Sol Invictus; 25: Sol Invictus; 26TL: Robert Lackenback/Time Life Pictures/Getty Images; 26 TR: Carl Mydans/Time Life Pictures/Getty Images; 26 BL: Fox Photos/Getty Images; 26 BR: Library of Congress/Getty Images; 27 TR: Paul Schutzer/Time Life Pictures/Getty Images; 27 TL: Central Press/Getty Images; 27 M: Libor Hajsky/AFP/Getty Images; 27 B: Bill Eppridge/Time Life Pictures/Getty Images; 26: Sol Invictus/NASA; 29: Sol Invictus/NASA; 30: ESA/NASA/SOHO/EIT; 32: Sol Invictus/NASA; 33: Sol Invictus/NASA; 34: Sol Invictus/NASA; 35: Sol Invictus/NASA; 36: Sol Invictus/NASA; 37: Sol Invictus/NASA; 41: Sol Invictus/NASA; 42: Sol Invictus/NASA; 43: Sol Invictus/NASA; 44: NASA; 46: Sol Invictus/NASA; 47: Sol Invictus/NASA; 48: Sol Invictus/NASA; 49: Sol Invictus/NASA; 50 TL: Keystone/Getty Images; 50 TR: Ossinger/dpa/Corbis; 50 BL: EFE/Corbis; 50 BR: CBS Photo Archive/Getty Images; 51 L: Agence France Presse/Getty Images; 51 TR: Evening Standard/Getty Images; 51 M: Keystone/Getty Images; 51B: Karen Kasmauski/Corbis; 53: Sol Invictus/NASA; 54: Detlev van Ravenswaay/Science Photo Library; 56: Sol Invictus/NASA; 57: Sol Invictus/NASA; 58: Sol Invictus/NASA; 59: Sol Invictus/NASA; 60: Sol Invictus/NASA; 61: Sol Invictus/NASA; 62: Sol Invictus/NASA; 63: Sol Invictus/NASA; 64: Sol Invictus/NASA; 65: NASA; 67: Stefen Seip/NASA; 68: NASA; 69: Sol Invictus/NASA; 70: Novosti/Science Photo Library; 71: Sol Invictus; 72: NASA/JPL; 74: ESA/Virtis/INAF; 77: NASA; 78: Sol Invictus/NASA; 79: Sol Invictus/NASA; 80TL: Hulton Archive/Getty Images; 80 TR: Robert Maass/Corbis; 80 BL: Diana Walker/Liaison/Getty Images; 80 BR: Georges de Keerle/Getty Images; 81 TL: Laski Diffusion./Liaison/Getty Images; 81 TR: Dirck Halstead/Time Life Pictures/Getty Images; 81 B: Gerard Malie/AFP/Getty Images; 85: Science Photo Library; 86: Sol Invictus/NASA; 87: NASA; 88: Sol Invictus/NASA; 89: Sol Invictus/NASA; 90: Sol Invictus/NASA; 91: NASA; 93: NASA/JPL/UA; 94: ESA; 95: MSSS/NASA; 98–147: Sol Invictus; 149: NASA; 150: NASA; 151L: The Galileo Project/NASA; 151R: Hubble/P. Thomas/B. Zellner/NASA; 152: Johns Hopkins University/APL/NASA; 153: NASA; 154 L: Johns Hopkins University/APL/NASA; 154R: JAXA; 155: Stephen Ostro (JPL), Arecibo Radio Telescope/NSF/NASA; 156: NASA/ESA/H. Weaver (JHU/APL), A. Stern (SwRI) and HST Pluto Team; 158: Sol Invictus/NASA; 160: Sol Invictus/NASA; 161: Sol Invictus/NASA; 162: H. Hammel, MIT/NASA/ESA; 163 L: NASA/JPL; 163TM: NASA/JPL; 163TR: NASA./JPL/University of Arizona; 163 BR: NASA; 164: NASA/JPL; 166T: NASA; 166M: Cassini Imaging Team/SSI/JPL/ESA/NASA; 166B: NASA/JPL/Space Science Institute; 167: NASA/JPL/USGS; 168T: Lawrence Sromovsky, UW Madison Space Science and Engineering Center; 169: NASA; 169: NASA/JPL-Caltech; 170: NASA; 172 TL: Alexander Joe/AFP/Getty Images; 172 TR: DOD/Time Life Pictures/Getty Images; 172 BL: Reuters/Corbis; 172 BR: Tim Graham/Corbis; 173TL: Louise Gubb/Corbis SABA; 173 TR: Stephen Ferry Liaison/Getty Images; 173 M: George Bridges/AFP/Getty Images; 173 B: Google; 175: NASA/JPL-Caltech; 176: Sol Invictus; 177 R: Alamy; 177 L: Sol Invictus; 178: ESA; 179 T: Science Photo Library; 179 B: ESO; 181: Charles and Josette Lenars/Corbis; 183: Roger Ressmeyer/Corbis; 184: N A Sharp, NOAO/NSO/Kitt Peak FTS/AURA/NSF; 185: NASA/ESA; 187: NASA/ESA/JPL/Arizona State University; 188: NASA/JPL-Caltech/University of Toledo/NOAO; 191: NASA/ESA/STScl/ Arizona State University; 193: NASA/JPL-Caltech; 194: ESA; 195: NASA/Hubble Heritage Team; 196 T: NASA/STScl/ACS Science Team/ESA; 196 B: NASA/JPL-Caltech; 197TL: NASA/Hubble Heritage Team; 197 TR: NASA/Hubble Heritage Team; 197B: NASA/Hubble Heritage Team; 198: Andrew Fruchter (STScl)/NASA; 199: NASA/JPL-Caltech/STScl; 200: Courtesy of ESA; 203: NASA/ESA/HUDF Team; 204 TL: Ber Murphy/Timepix/Time Life Pictures/Getty Images; 204 TR: James Nielsen/AFP/Getty Images; 204 BL: Deutsche Bundesbank/Getty Images; 204 BR: CNN/Getty Images; 204 TL: John Russell/AFP/Getty Images; 204 TR: Reinhard Krause/Reuters/Corbis; 204 B: NOAA/Getty Images; 207: NASA/KSC; 208: JAC/UCLA; 209: NASA; 210: NASA/James Bell/Michael Wolff/Hubble; 211: Sergei Chirikov/EPA/Corbis.